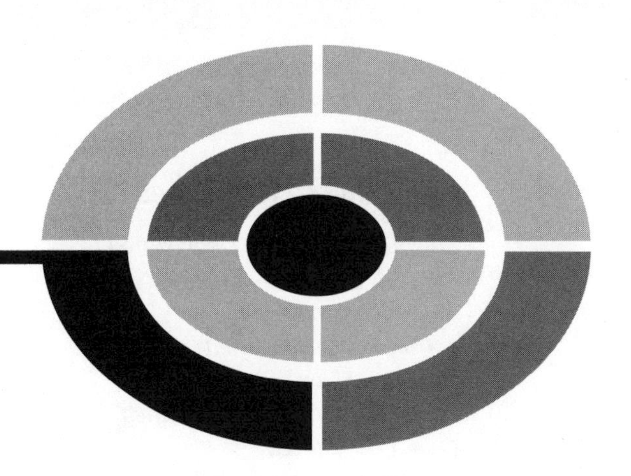

METEOROLOGY DEMYSTIFIED

Demystified Series

Advanced Statistics Demystified
Algebra Demystified
Anatomy Demystified
Astronomy Demystified
Biology Demystified
Business Statistics Demystified
C++ Demystified
Calculus Demystified
Chemistry Demystified
College Algebra Demystified
Data Structures Demystified
Databases Demystified
Differential Equations Demystified
Digital Electronics Demystified
Earth Science Demystified
Electricity Demystified
Electronics Demystified
Environmental Science Demystified
Everyday Math Demystified
Geometry Demystified
Home Networking Demystified
Investing Demystified
Java Demystified
JavaScript Demystified
Macroeconomics Demystified
Math Proofs Demystified
Math Word Problems Demystified
Microbiology Demystified
OOP Demystified
Options Demystified
Personal Computing Demystified
Physics Demystified
Physiology Demystified
Pre-Algebra Demystified
Precalculus Demystified
Probability Demystified
Project Management Demystified
Quantum Mechanics Demystified
Relativity Demystified
Robotics Demystified
Six Sigma Demystified
Statistics Demystified
Trigonometry Demystified

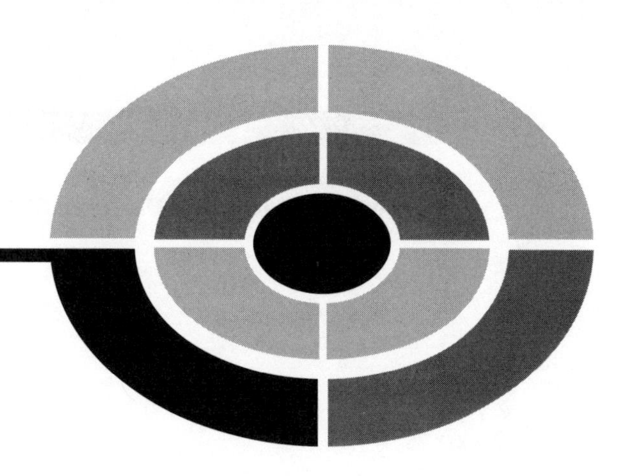

METEOROLOGY DEMYSTIFIED

STAN GIBILISCO

McGRAW-HILL
New York Chicago San Francisco Lisbon London
Madrid Mexico City Milan New Delhi San Juan
Seoul Singapore Sydney Toronto

The McGraw-Hill Companies

Cataloging-in-Publication Data is on file with the Library of Congress

1 2 3 4 5 6 7 8 9 0 DOC/DOC 0 1 0 9 8 7 6 5

ISBN 0-07-144848-9

The sponsoring editor for this book was Judy Bass and the production supervisor was Pamela A. Pelton. It was set in Times Roman by Fine Composition. The art director for the cover was Margaret Webster-Shapiro.

Printed and bound by RR Donnelley.

This book is printed on recycled, acid-free paper containing a minimum of 50% recycled, de-inked fiber.

McGraw-Hill books are available at special quantity discounts to use as premiums and sales promotions, or for use in corporate training programs. For more information, please write to the Director of Special Sales, McGraw-Hill Professional, McGraw-Hill, Two Penn Plaza, New York, NY 10121-2298. Or contact your local bookstore.

To Samuel, Tim, and Tony
from Uncle Stan

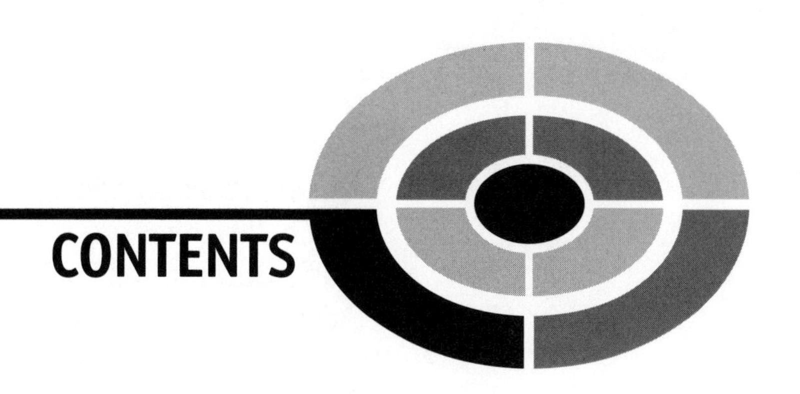

CONTENTS

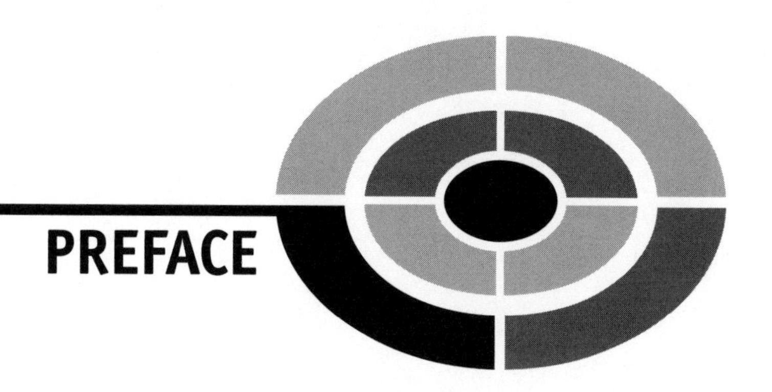

PREFACE

This book is for people who want to learn the fundamentals of meteorology without taking a formal course. It can serve as a supplemental text in a classroom, tutored, or home-schooling environment. I recommend that you start at the beginning of this book and go straight through.

There are "conversational" problems and solutions scattered throughout the text. There is a practice quiz at the end of each chapter, and a final exam at the end of the book. The quiz and exam questions are multiple-choice, and are similar to the sorts of questions used in standardized tests.

The chapter-ending quizzes are "open-book." You may (and should) refer to the chapter texts when taking them. When you think you're ready, take the quiz, write down your answers, and then give your list of answers to a friend. Have the friend tell you your score, but not which questions you got wrong. Stick with a chapter until you get most (hopefully all) of the answers right. The correct choices are listed in the appendix.

Take the final exam when you have finished all the chapters and chapter-ending quizzes. A satisfactory score is at least 75 correct answers. With the final exam, as with the quizzes, have a friend tell you your score without letting you know which questions you missed. Note the exam topics (if any) that give you trouble. Review those topics in the text. Then take the exam again, and see if you get a better score.

I recommend that you complete one chapter a week. An hour or two daily ought to be enough time for this. That way, you'll complete the course in a little over two months. When you're done with the course, you can use this book, with its comprehensive index, as a permanent reference.

Suggestions for future editions are welcome.

STAN GIBILISCO

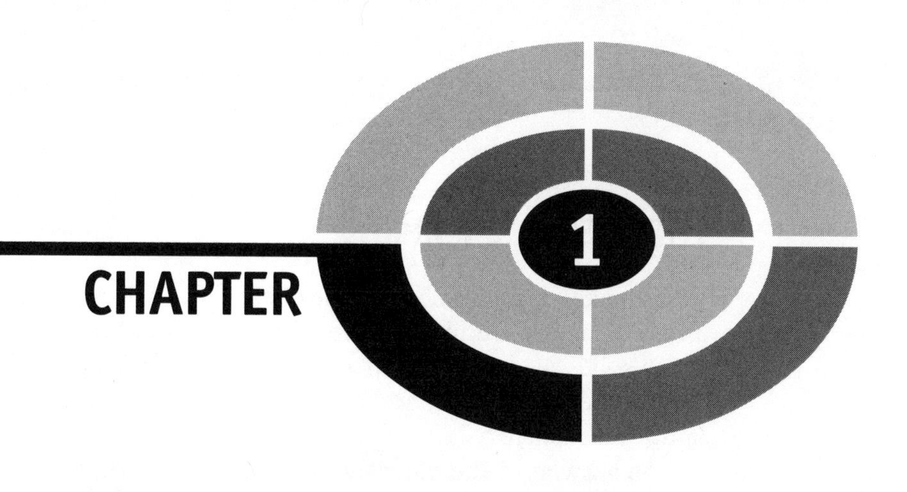

Background Physics

Thousands of years ago, the alchemists believed that all things in the material universe consist of combinations of four "elements": *earth*, *water*, *air*, and *fire*. According to this theory, different proportions of these four "elements" give materials their unique properties. Later, physical scientists discovered that there are dozens of elements, and even these are not the fundamental constituents of matter. Three basic *states of matter* are recognized by scientists today. These states, also called *phases*, are known as the *solid phase* (the latter-day analog of earth), the *liquid phase* (the analog of water), and the *gaseous phase* (the analog of air). A sample of matter in one of these states is called a *solid*, a *liquid*, or a *gas*.

The Solid Phase

A sample of matter in the solid phase retains its shape unless it is subjected to violent impact, placed under stress, or put in an environment with extremely high temperature. Examples of solids at room temperature are rock, salt, wood, and plastic.

THE ELECTRIC FORCE

What makes a solid behave as it does? After all, we've all been told that the *atoms* of matter are mostly empty space, and that this is true even in the most dense solids we see on this planet. So why can't solid objects pass through one another the way galaxies sometimes do in outer space, or the way dust clouds do in the atmosphere?

The answers to these questions can be found when we analyze the electrical forces in and around atoms. Every atom consists of a small, dense, positively charged *nucleus*, orbited by negatively charged *electrons* that follow mean (average) paths called *shells*. Objects with electrical charges of the same polarity (negative–negative or positive–positive) always repel each other. The closer together two objects with the same type of charge come to each other, the more forcefully they repel. Thus, even when an atom has an equal number of electrons and protons so it is electrically neutral as a whole, the charges are concentrated in different places. The positive charge is contained in the nucleus, and the negative charge surrounds the nucleus in one or more shells. These shells are usually shaped like concentric spheres.

Suppose you could shrink down to submicroscopic size and stand on the surface of a sheet of an elemental metal such as aluminum or copper. Below you, the surface would appear like a large, flat field full of rigid spheres (Fig. 1-1). You would find the spheres resistant to penetration by other spheres. All the spheres

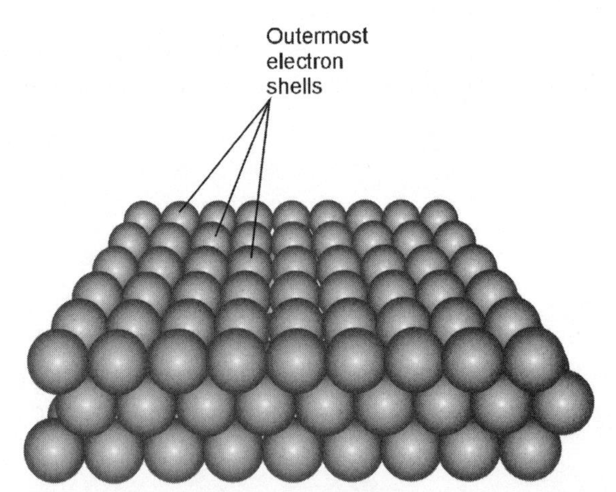

Fig. 1-1. In a solid, the outer electron shells of the atoms are tightly packed. This drawing is greatly oversimplified.

would be negatively charged, so they would all repel each other. This would keep them from passing through each other, and would also keep the surface in a stable, fixed state. The spheres would be mostly empty space inside, but there wouldn't be much space in between them. They would be tightly packed together.

The foregoing is an oversimplification, but it should give you an idea of the reason why solids don't normally pass through each other, and why many solids resist penetration even by liquids such as water, or by gases such as air.

DENSITY OF SOLIDS

The *density* of a solid is measured in terms of the number of *kilograms* (kg) per *cubic meter* (m^3). That is, density is equal to *mass* divided by *volume*. The *kilogram per meter cubed* (kg/m^3 or $kg \times m^{-3}$) is the measure of density in the *International System of units* (SI), also known as the *meter-kilogram-second* (mks) system. This is a rather awkward unit in most real-life situations. Imagine trying to determine the density of sandstone by taking a cubical chunk of the stuff measuring one meter (1 m) on an edge, and placing it on a laboratory scale! You'd need a construction crane to lift the boulder, and it would smash the scale.

Because of the impracticality of measuring density directly in standard international units, the *centimeter-gram-second* (cgs) unit is sometimes used instead. This is the number of *grams* of mass (g) per *cubic centimeter* (cm^3) of the material in question. Technically it is called the *gram per centimeter cubed* (g/cm^3 or $g \times cm^{-3}$). To convert the density of a given sample from grams per centimeter cubed to kilograms per meter cubed, multiply by 1000 (10^3). To convert the density of a sample from kilograms per meter cubed to grams per centimeter cubed, multiply by 0.001 (10^{-3}).

You can think of solids that are dense, such as lead. Iron is dense, too. Aluminum is not as dense. Rocks are less dense than most metals. Glass has about the same density as silicate rock, from which it is made. Wood, and most plastics, are not very dense.

MEASURING SOLID VOLUME

Samples of solids rarely come in perfect blocks, cubes, or spheres, which are shapes that lend themselves to calculation of volume by mathematical formulas. Most samples are irregular, and defy direct dimensional measurement.

Scientists have an indirect way of measuring the volumes of irregular solid samples: immerse them in a liquid. First, we measure the amount of liquid in a

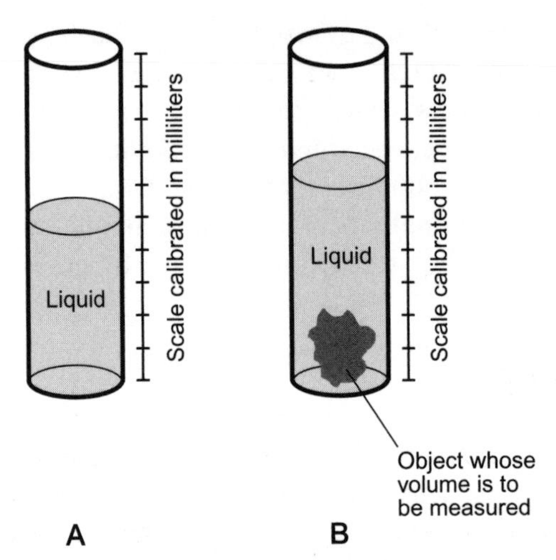

Fig. 1-2. Measuring the volume of a solid. At A, container with liquid but without the sample; at B, container with the sample totally submerged in the liquid.

container (Fig. 1-2A). Then we measure the amount of liquid that is displaced when the object is completely submerged. This shows up as an increase in the apparent amount of liquid in the container (Fig. 1-2B). One *milliliter* (1 ml) of liquid volume happens to be exactly equal to 1 cm^3, and any good chemist has a few containers that are marked off in milliliters. That's the way to do it, then—provided the solid does not dissolve in the liquid, none of the liquid is absorbed into the solid, and the liquid doesn't evaporate too fast.

SPECIFIC GRAVITY OF SOLIDS

Another important characteristic of a solid is its density relative to that of pure liquid water at 4 degrees Celsius (4°C), which is about 39 degrees Fahrenheit (39°F). Water attains its greatest density at this temperature, and in this condition it is assigned a *relative density* of 1. Liquid water at 4°C has a density of approximately 1000 kg/m^3, which is equal to 1 g/cm^3. Substances with relative density greater than 1 sink in pure water at 4°C, and substances with relative density less than 1 float in pure water at 4°C. The relative density of a solid, defined in this way, is called the *specific gravity*, abbreviated as sp gr.

You can think of substances whose specific gravity numbers are greater than 1. Examples include most rocks and virtually all metals. But *pumice*, a volcanic rock that is filled with air pockets, floats on water. Most of the planets, their moons, and the asteroids and meteorites in our Solar System have specific gravity greater than 1, with the exception of Saturn, which would float if a lake big enough could be found in which to test it!

Interestingly, water ice has specific gravity less than 1 (sp gr < 1), so it floats on liquid water. This property of ice allows fish to live underneath the frozen surfaces of lakes in the winter in the temperate and polar regions. The surface layer of ice acts as an insulator against the cold atmosphere. If ice had a specific gravity larger than 1 (sp gr > 1), it would sink to the bottoms of lakes during the winter months. This would leave the surfaces constantly exposed to temperatures below freezing, causing more and more of the water to freeze, until shallow lakes would become frozen from the surface all the way to the bottom. In such an environment, all the fish would die during the winter, because they wouldn't be able to extract the oxygen they need from the solid ice, nor would they be able to swim around in order to feed themselves. Many other aquatic creatures would be similarly affected.

PROBLEM 1-1

A sample of solid matter has a volume of 45.3 cm^3 and a mass of 0.543 kg. What is the density in grams per centimeter cubed?

SOLUTION 1-1

This problem is a little tricky, because two different systems of units are used: SI for the volume and cgs for the mass. To get a meaningful answer, we must be consistent with our units. The problem requires that we express the answer in the cgs system, so let's convert kilograms to grams. This means we have to multiply the mass figure by 1000, which tells us that the sample masses 543 g. Determining the density in grams per centimeter cubed is now a simple arithmetic problem: divide the mass by the volume. If d is density, m is mass, and V is volume, then they are related by the following formula:

$$d = m/V$$

In this case:

$$d = 543/45.3 = 12.0 \text{ g/cm}^3$$

This answer is rounded to three *significant figures*.

PROBLEM 1-2

Calculate the density of the sample from Problem 1-1 in kilograms per meter cubed. Do not use the conversion factor on the result of Problem 1-1. Start from scratch.

SOLUTION 1-2

This requires that we convert the volume to units in SI, that is, to meters cubed. There are 1,000,000, or 10^6, centimeters cubed in a meter cubed. Therefore, in order to convert this cgs volume to volume in SI, we must divide by 10^6, the equivalent of multiplying by 10^{-6}. This gives us 45.3×10^{-6} m^3, or 4.53×10^{-5} m^3 in standard scientific notation, as the volume of the object. Now we can divide the mass by the volume directly:

$$\begin{aligned} d &= m/V \\ &= 0.543/(4.53 \times 10^{-5}) \\ &= 0.120 \times 10^5 \\ &= 1.20 \times 10^4 \text{ kg/m}^3 \end{aligned}$$

This is rounded to three significant figures.

The Liquid Phase

In the liquid state or phase, matter has two properties that distinguish it from matter in the solid phase. First, a liquid changes shape so that it conforms to the inside boundaries of any container in which it is placed. Second, a liquid placed in an open container (such as a jar or bucket) flows to the bottom of the container and develops a defined, flat surface in an environment where there is constant force caused by gravitation or acceleration.

DIFFUSION OF LIQUIDS

Imagine a jar on board a space ship in which the environment is weightless (there is no force caused by gravitation or acceleration). Suppose that the jar is filled with liquid, and then another liquid that does not react chemically with the first liquid is introduced into the jar. Gradually, the two liquids blend together until the mixture is uniform throughout the jar. This blending process is called *diffusion*.

Some pairs of liquids undergo the diffusion process more readily than others. Alcohol diffuses into water at room temperature in a short time. But heavy motor oil diffuses into light motor oil slowly, and motor oil hardly diffuses into water at all. When two liquids readily diffuse into one another, the process happens without the need for shaking the container, because the atoms of a liquid are always in motion, and this motion causes them to jostle each other until they become uniformly mixed.

If the same experiment is conducted in a bucket on the surface of the earth, where there is gravitational force, diffusion occurs, but "heavier" (more dense) liquids tend to sink towards the bottom and "lighter" (less dense) liquids tend to rise toward the surface. Alcohol, for example, "floats" on water. But the boundary between the alcohol and water is not sharply defined, as is the surface between the water and the air. The motion of the atoms constantly "tries" to mix the two liquids.

VISCOSITY OF LIQUIDS

Some liquids flow more easily than others. You know there is a difference at room temperature between, say, water and thick molasses. If you fill a glass with water and another glass with an equal amount of molasses and then pour the contents of both glasses into the sink, the glass containing the water will empty much faster. The molasses is said to have higher *viscosity* than the water at room temperature. On an extremely hot day, the difference is less obvious than it is on a cold day.

Some liquids are far more viscous even than thick molasses. An example of a liquid with extremely high viscosity is asphalt, as it is poured to make the surface of a new highway. Another example is petroleum jelly. These substances meet the criteria as defined above to qualify as liquids, but they are thick. As the temperature goes down, these substances become less like liquids and more like solids. It is impossible to draw an exact line between the liquid and the solid phases for either of these two substances.

LIQUID OR SOLID?

There is not always a specific answer to the question, "Is this substance a solid or a liquid?" It often depends on the observer's point of reference. Some substances can be considered solid in the short-term time sense, but liquid in the long-term sense. An example is the *mantle* of the earth, the layer of rock between the crust and the core. In a long-term time sense, pieces of the crust, known as *tectonic plates*, float around on top of the mantle like scum on the surface of a hot vat of liquid. This is manifested as *plate tectonics*, which used to be known as "continental drift." It is apparent when the earth's history is evaluated over periods of millions of years. But from one moment (as we perceive it) to the next, and even from hour to hour or from day to day, the crust seems rigidly fixed to the mantle. The mantle therefore behaves like a solid in short-term time frames, but like a liquid in long-term time frames.

Imagine that we could turn ourselves into creatures whose life spans were measured in trillions (units of 10^{12}) of years, so that 1,000,000 years seemed to pass like a moment. Then from our point of view, the earth's mantle would behave as a liquid with low viscosity, just as water seems to us in our actual state of time awareness. If we could become creatures whose entire lives lasted only a tiny fraction of a second, then liquid water would seem to take eons to get out of a glass tipped on its side, and we would conclude that this substance was solid, or perhaps a liquid with high viscosity.

DENSITY OF LIQUIDS

The density of a liquid can be defined in three ways. The *mass density of a liquid* is defined in terms of the number of kilograms per meter cubed (kg/m^3) in a sample of liquid. The *weight density of a liquid* is defined in *newtons per meter cubed* (N/m^3 or $N \times m^{-3}$), and is equal to the mass density multiplied by the acceleration in *meters per second squared* (m/s^2 or $m \times s^{-2}$) to which the sample is subjected. The *particle density of a liquid* is defined as the number of *moles per meter cubed* (mol/m^3 or $mol \times m^{-3}$), where 1 mole represents approximately 6.02×10^{23} atoms.

Let d_m be the mass density of a liquid sample (in kilograms per meter cubed), let d_w be the weight density (in newtons per meter cubed), and let d_p be the particle density (in moles per meter cubed). Let m represent the mass of the sample (in kilograms), let V represent the volume of the sample (in meters cubed), and let N represent the number of moles of atoms in the sample. Let a be the acceleration (in meters per second squared) to which the sample is subjected. Then the following equations hold:

$$d_m = m/V$$
$$d_w = ma/V$$
$$d_p = N/V$$

Note the difference here between the non-italic uppercase N, which represents newtons, and the italic uppercase *N*, which represents the number of moles of atoms in a sample.

Alternative definitions for mass density, weight density, and particle density use the *liter*, which is equal to a thousand centimeters cubed (1000 cm^3) or one-thousandth of a meter cubed (0.001 m^3), as the standard unit of volume. Once in awhile you'll see the centimeter cubed (cm^3), also known as the *milliliter* because it is equal to 0.001 liter, used as the standard unit of volume.

These are simplified definitions, because they assume that the density of the liquid is uniform throughout the sample.

PROBLEM 1-3

A sample of liquid measures 0.2750 m^3. Its mass is 300.0 kg. What is its mass density in kilograms per meter cubed?

SOLUTION 1-3

This is straightforward, because the input quantities are already given in SI. There is no need for us to convert from grams to kilograms, from milliliters to meters cubed, or anything like that. We can simply divide the mass m by the volume V, as follows:

$$\begin{aligned} d_{\mathrm{m}} &= m/V \\ &= 300.0 \text{ kg}/0.2750 \text{ m}^3 \\ &= 1090 \text{ kg/m}^3 \end{aligned}$$

We're entitled to go to four significant figures here, because our input numbers are both given to four significant figures.

PROBLEM 1-4

Given that the acceleration of gravity at the earth's surface is 9.81 m/s^2, what is the weight density of the sample of liquid described in Problem 1-4?

SOLUTION 1-4

All we need to do in this case is multiply our mass density answer by 9.81 m/s^2. This gives us:

$$\begin{aligned} d_{\mathrm{w}} &= 1090 \text{ kg/m}^3 \times 9.81 \text{ m/s}^2 \\ &= 10{,}700 \text{ N/m}^3 = 1.07 \times 10^4 \text{ N/m}^3 \end{aligned}$$

In this case, we can go to only three significant figures, because that is the extent of the precision with which the acceleration of gravity is specified.

MEASURING LIQUID VOLUME

The volume of a liquid sample is usually measured using a test tube or flask marked off in milliliters or liters. But there's another way to measure the volume of a liquid sample, provided we know its chemical composition and the weight density of the substance in question. That is to weigh the sample of liquid, and then divide the weight by the weight density. We must, of course, pay careful attention to the units. In particular, the weight must be expressed in newtons, which is equal to the mass in kilograms times the acceleration of gravity (9.81 m/s^2).

Let's do a mathematical exercise to show how we can measure volume in this way. Let d_w be the known weight density of a huge sample of liquid, too large for its volume to be measured using a flask or test tube. Suppose this substance has a weight of w, expressed in newtons. If V is the volume in meters cubed, we know from the formula for weight density that:

$$d_w = w/V$$

because $w = ma$, where m is the mass in kilograms, and a is the acceleration of gravity in meters per second squared. If we divide both sides of this equation by w, we get:

$$d_w/w = 1/V$$

Then we can invert both sides of this equation, and exchange the left-hand and the right-hand sides, to obtain:

$$V = w/d_w$$

This is based on the assumption that V, w, and d_w are all nonzero quantities. This is always true in the real world; all materials occupy at least some volume, have at least some weight because of gravitation, and have nonzero density because there is always some "stuff" in a finite amount of physical space.

PRESSURE IN LIQUIDS

Have you read, or been told, that liquid water can't be compressed? In a simplistic sense, that is true, but this doesn't mean liquid water never exerts pressure.

Liquids can and do exert pressure, as anyone who has been in a flash flood or a hurricane, or who has gone deep-sea diving, will tell you. You can experience "water pressure" for yourself by diving down several feet in a swimming pool and noting the sensation the water produces as it presses against your eardrums.

In a fluid, the pressure, which is defined in terms of force per unit area, is directly proportional to the depth. Pressure is also directly proportional to the weight density of the liquid. Let d_w be the weight density of a liquid (in newtons per meter cubed), and s be the depth below the surface (in meters). Then the pressure, P (in newtons per meter squared) exerted by the liquid at that depth is given by:

$$P = d_w s$$

If we are given the mass density d_m (in kilograms per meter cubed) rather than the weight density, the formula becomes:

$$P = 9.81 d_m s$$

PROBLEM 1-5

Liquid water at room temperature has a mass density of 1000 kg/m^3. How much force is exerted on the outer surface of a cube measuring 10.000 cm on an edge, submerged 1.00 m below the surface of a body of water?

SOLUTION 1-5

First, figure out the total surface area of the cube. It measures 10.000 cm, or 0.10000 m, on an edge, so the surface area of one face is 0.10000 m $\times$ 0.10000 m = 0.010000 m^2. There are six faces on a cube, so the total surface area of the object is 0.010000 m^2 $\times$ 6 = 0.060000 m^2. (Don't be irritated by the "extra" zeroes here. They indicate that the length of the edge of the cube has been specified to five significant figures. Also, the number 6 is an exact quantity, so it can be considered accurate to as many significant figures as we want.)

Next, figure out the weight density of water (in newtons per meter cubed). This is 9.81 times the mass density, or 9810 N/m^3. This is best stated as 9.81×10^3 N/m^3, because we are given the acceleration of gravity to only three significant figures, and scientific notation makes this fact clear. (From this point on let's stick with power-of-10 notation so we don't make the mistake of accidentally claiming more accuracy than that to which we're entitled.)

The cube is at a depth of 1.00 m, so the water pressure at that depth is 9.81×10^3 N/m^3 $\times$ 1.00 m = 9.81×10^3 N/m^2. The force F (in newtons) on the cube is therefore equal to this number multiplied by the surface area of the cube:

$$F = 9.81 \times 10^3 \text{ N/m}^2 \times 6.00000 \times 10^{-2} \text{ m}^2$$
$$= 58.9 \times 10^1 \text{ N} = 589 \text{ N}$$

PASCAL'S LAW FOR INCOMPRESSIBLE LIQUIDS

Imagine a watertight, rigid container. Suppose there are two pipes of unequal diameters running upwards out of this container. Imagine that you fill the container with an incompressible liquid such as water, so the container is completely full and the water rises partway up into the pipes. Suppose you place pistons in the pipes so they make perfect water seals, and then you leave the pistons to rest on the water surface (Fig. 1-3).

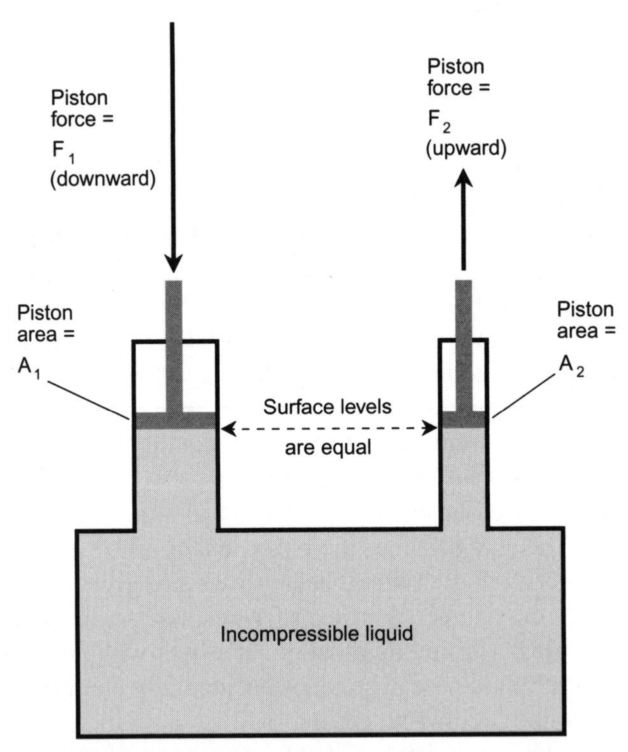

Fig. 1-3. Pascal's law for confined, incompressible liquids. The forces are directly proportional to the surface areas where the pistons contact the liquid.

Because the pipes have unequal diameters, the surface areas of the pistons are different. One of the pistons has area A_1 (in meters squared), and the other has area A_2 (also in meters squared). Suppose you push downward on piston number 1 (the one whose area is A_1) with a force F_1 (in newtons). How much upward force, F_2, is produced at piston number 2 (the one whose area is A_2)? *Pascal's law* provides the answer: the forces are directly proportional to the areas of the piston faces in terms of their contact with the liquid. In the example shown by Fig. 1-3, piston number 2 is smaller than piston number 1, so the force F_2 is proportionately less than the force F_1. Mathematically, the following equations both hold:

$$F_1/F_2 = A_1/A_2$$
$$A_1F_2 = A_2F_1$$

When using either of these equations, we must be consistent with units throughout the calculations. Also, the top equation is meaningful only as long as the force exerted is nonzero.

PROBLEM 1-6

Suppose the areas of the pistons shown in Fig. 1-3 are $A_1 = 12.00\ \text{cm}^2$ and $A_2 = 15.00\ \text{cm}^2$. (This does not seem to agree with the illustration, where piston number 2 looks smaller than piston number 1, but forget about that while we solve this problem.) If you press down on piston number 1 with a force of 10.00 N, how much upward force results at piston number 2?

SOLUTION 1-6

At first, you might think we have to convert the areas of the pistons to meters squared in order to solve this problem. But in this case, it is sufficient to find the ratio of the areas of the pistons, because both areas are given to us in the same units:

$$A_1/A_2 = 12.00\ \text{cm}^2 / 15.00\ \text{cm}^2$$
$$= 0.8000$$

Thus, we know that $F_1/F_2 = 0.8000$. We are given $F_1 = 10.00$ N, so it is easy to solve for F_2:

$$10.00/F_2 = 0.8000$$
$$1/F_2 = 0.08000$$
$$F_2 = 1/0.08000 = 12.50\ \text{N}$$

We are entitled to four significant figures throughout this calculation, because all the input data is provided to that degree of precision.

The Gaseous Phase

The gaseous phase of matter is similar to the liquid phase, insofar as a gas conforms to the boundaries of a container or enclosure. But a gas is much less affected by gravity than a liquid on a small scale. If you fill up a bottle with a gas, there is no discernible "surface" to the gas. Another difference between liquids and gases is the fact that gases are nearly always compressible.

GAS DENSITY

The density of a gas can be defined in three ways, exactly after the fashion of liquids. The *mass density of a gas* is defined in terms of the number of kilograms per meter cubed (kg/m^3) that a sample of gas has. The *weight density of a gas* is defined in newtons per meter cubed (N/m^3), and is equal to the mass density multiplied by the acceleration in meters per second squared (m/s^2) to which the sample is subjected. The *particle density of a gas* is defined as the number of moles of atoms per meter cubed (mol/m^3) in a parcel or sample of gas, where $1 \text{ mol} \approx 6.02 \times 10^{23}$.

DIFFUSION IN SMALL CONTAINERS

Imagine a rigid enclosure, such as a glass jar, from which all of the air has been pumped. Suppose this jar is placed somewhere out in space, far away from the gravitational effects of stars and planets, and where space itself is a near vacuum (compared to conditions on the surface of the earth, anyhow). Suppose the temperature is the same as that in a typical household. Now suppose a certain amount of elemental gas is pumped into the jar. The gas distributes itself quickly throughout the interior of the jar.

Now suppose another gas that does not react chemically with the first gas is introduced into the chamber to mix with the first gas. The diffusion process occurs rapidly, so the mixture is uniform throughout the enclosure after a short time. It happens so fast because the atoms in a gas move around furiously, often colliding with each other, and their motion is so energetic that they spread out inside any container of reasonable size (Fig. 1-4A).

If the same experiment were performed in the presence of a gravitational field, the gases would still mix uniformly inside the jar. This happens with nearly all gases in containers of small size. Inside a large enclosure such as a gymnasium, gases with significantly greater mass density than that of the air in general

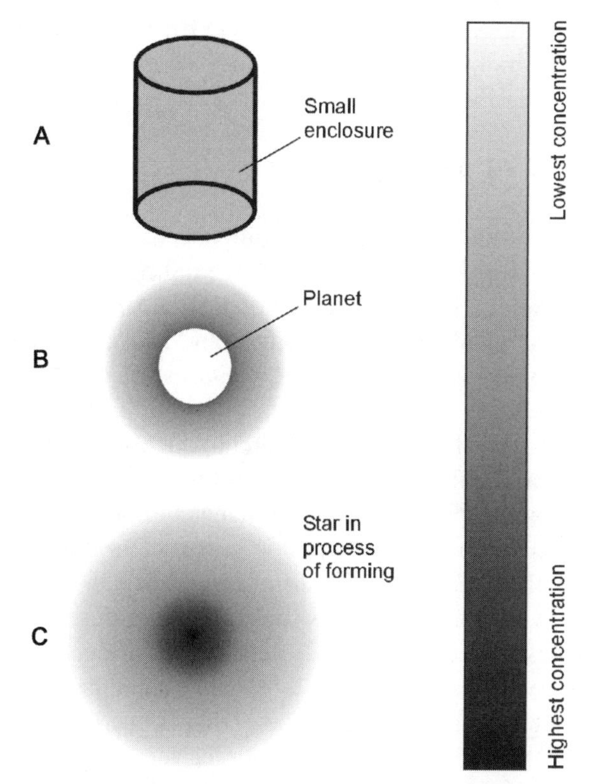

Fig. 1-4. At A, distribution of gas inside a container. At B, distribution of gas around a planet with an atmosphere. At C, distribution of gas in a star as it is forming. Darkest shading indicates highest concentration.

tend to "sink to the bottom" if there is poor ventilation and poor air circulation. This results in greater concentrations of especially heavy gases near floors than near the ceilings. However, a little air movement (such as can be provided by fans) will cause even the heavier gases to quickly diffuse uniformly throughout the enclosure.

Planetary atmospheres, such as that of our own earth, consist of mixtures of various gases. In the case of our planet, approximately 78% of the gas in the atmosphere at the surface is nitrogen, 21% is oxygen, and the remaining 1% consists of many other gases, including argon, carbon dioxide, carbon monoxide, hydrogen, helium, ozone (oxygen molecules with three atoms rather than the usual two), and tiny quantities of some gases that would be poisonous in high concentrations, such as chlorine and methane. These gases blend uniformly in containers or enclosures of reasonable size, even though some of them have atoms that are far more massive than others. Diffusion ensures this, as long as there is the slightest bit of air movement.

GASES NEAR A PLANET

Now imagine the shroud of gases that compose the atmosphere of a planet. Gravitation attracts some gas from the surrounding space. Other gases are ejected from the planet's interior during volcanic activity. Still other gases are produced by the biological activities of plants and animals, if the planet harbors life. On the earth, some gases are produced by industrial activity and by the combustion of fossil fuels.

All the gases in the earth's atmosphere tend to diffuse into each other when we look at parcels of reasonable size, regardless of the altitude above the surface. But there is unlimited "outer space" around our planet and only a finite amount of gas near its surface, and the gravitational pull is greater near the surface than far out in space. Because of these factors, diffusion takes place in a different way when considered all the way from the earth's surface up into outer space. The greatest concentration of gas molecules (particle density) occurs near the surface, and it decreases with increasing altitude (Fig. 1-4B). The same is true of the number of kilograms per meter cubed of the atmosphere, that is, the mass density of the gas.

On the large scale of the earth's atmosphere, yet another effect takes place. For a given number of atoms or molecules per meter cubed, some gases are more massive than others. Hydrogen is the least massive. Helium has low mass, too. Oxygen is more massive, and carbon dioxide more massive still. The most massive gases tend to sink toward the surface, while the least massive gases rise up high, and some of their atoms escape into outer space or are not permanently captured by the earth's gravitation.

There are no distinct boundaries, or layers, from one type of gas to another in the atmosphere. Instead, the transitions are gradual. That's good, because if the gases of the atmosphere were stratified in a defined way, we would have no oxygen down here on the surface. Instead, we'd be smothered in some noxious gas such as carbon dioxide or sulfur dioxide. We'd have to climb mountains in order to breathe!

GASES IN OUTER SPACE

Outer space was once believed to be a perfect vacuum. But this is not the case. There is plenty of gaseous matter out there, and much of it is hydrogen and helium. (There are also trace amounts of heavier gases, and plenty of solid rocks and ice chunks as well.) All the atoms in outer space interact gravitationally with all the others.

The motion of atoms in outer space is almost random, but not quite. The slightest perturbation in the randomness of the motion gives gravitation a chance to cause the gas to "clump" into huge clouds. Once this process begins, it can continue until a globe of gas forms in which the central particle density is significant (Fig. 1-4C). As gravitation continues to pull the atoms in toward the center, the mutual attraction among the atoms there becomes greater and greater.

If a gas cloud in space has some *spin*, it flattens into an oblate spherical shape and eventually into a disk with a bulge at the center. A vicious circle ensues, and the density in the central region skyrockets. The gas pressure in the center therefore rises, and this causes it to heat up. Ultimately it gets so hot that *nuclear fusion* begins, and a star is born. Similar events among the atoms of the gas on a smaller scale can result in the formation of asteroids, planets, and planetary moons.

GAS PRESSURE

Unlike most liquids, gases can be compressed. This is why it is possible to fill up hundreds of balloons with a single, small tank of helium gas, and why it is possible for a scuba diver to breathe for a long time from a single small tank of air.

Imagine a container whose volume (in meters cubed) is equal to V. Suppose there are N moles of atoms of a particular gas inside this container, which is surrounded by a perfect vacuum. We can say certain things about the pressure P, in newtons per meter squared, that the gas exerts outward on the walls of the container. First, P is proportional to N, provided that V is held constant. Second, if V increases while N remains constant, P decreases.

There is another important factor—temperature, symbolized T—that affects gases when the containers holding them expand or contract. When a parcel of gas is compressed, it heats up; when it is decompressed, it cools off. Heating up a parcel of gas increases the pressure, if all other factors are held constant, and cooling it off reduces the pressure.

What Is Heat?

Heat is a form of energy transfer that can occur between a given object, place, or region and another object, place, or region. For example, if you place a kettle of water on a hot stove, heat energy is transferred from the burner to the water.

This is *conductive heat*, also called *conduction* (Fig. 1-5A). When an infrared (IR) lamp, sometimes called a "heat lamp," shines on your sore shoulder, energy is transferred to your skin surface from the filament of the lamp; this is *radiative heat*, also called *radiation* (Fig. 1-5B). When a fan-type electric heater warms a room, air passes through the heating elements and is blown by a fan into the room where the heated air rises and mixes with the rest of the air in the room. This is *convective heat*, also called *convection* (Fig. 1-5C).

The definition of *heat* is not identical with the definition of *energy*, although heat and energy can both be defined in *joules* (symbolized J). The joule is the standard unit of energy in physics. Heat is the transfer of energy that occurs when conduction, radiation, and/or convection take place. Sometimes the energy transfer takes place in only one of these three modes, but sometimes it occurs in two or all three.

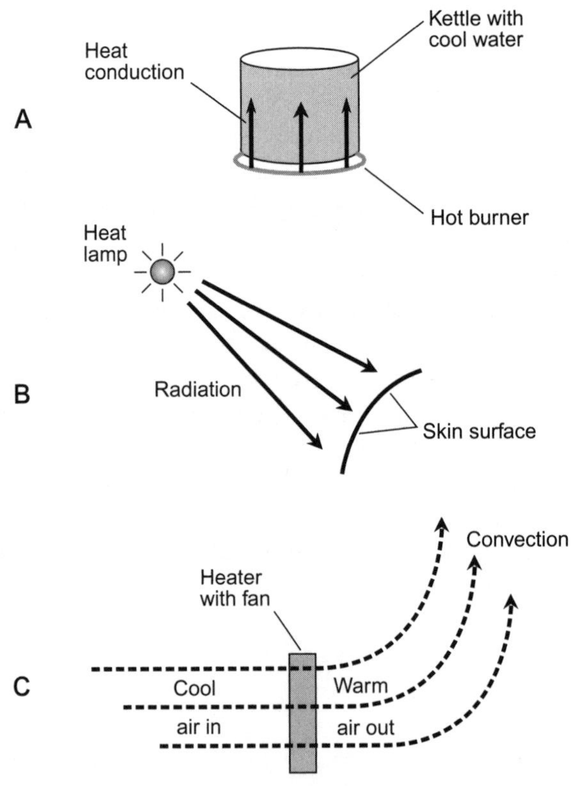

Fig. 1-5. Examples of heat energy transfer by conduction (A), radiation (B), and convection (C).

THE CALORIE

The *calorie* (symbolized cal) is an older unit of energy that is still used, especially in reference to heat transfer or exchange. One calorie is the equivalent of 4.184 J. In physics, the calorie is a much smaller unit than the "nutritionist's calorie," which is actually a physical *kilocalorie* (kcal), equal to 1000 cal.

The calorie (cal) in which we are interested is the amount of energy transfer that raises the temperature of exactly one gram (1 g) of pure liquid water by exactly one degree Celsius (1°C). It is also the amount of energy lost by 1 g of pure liquid water if its temperature falls by 1°C. The *kilocalorie* (kcal) is the amount of energy transfer involved when the temperature of exactly 1 kg, or 1000 g, of pure liquid water, rises or falls by exactly 1°C.

This definition of the calorie holds true only as long as the water is liquid during the entire process. If any of the water freezes, thaws, boils, or condenses, this definition is not valid. At standard atmospheric pressure on the earth's surface, in general, this definition holds for temperatures between approximately 0°C (the freezing point of water) and 100°C (the boiling point).

SPECIFIC HEAT

Pure liquid water requires one calorie per gram (1 cal/g) to warm it up or cool it down by 1°C (provided it is not at the melting/freezing temperature or the vaporization/condensation temperature, as we shall shortly see.) But what about oil or alcohol or salt water? What about solids such as steel or wood? What about gases such as air? It's not so simple then. A certain, fixed amount of heat energy raises or lowers the temperatures of fixed masses of some substances more than others. Some matter takes more than 1 cal/g to get hotter or cooler by 1°C; some matter takes less.

Suppose we have a sample of some liquid of unknown or unspecified chemical composition. Call it "substance X." We measure out one gram (1.00 g) of this liquid, accurate to three significant figures, by pouring some of it into a test tube placed on a laboratory balance. Then we transfer one calorie (1.00 cal) of energy to substance X. Suppose that, as a result of this energy transfer, this sample of substance X increases in temperature by 1.20°C? Obviously substance X is not water, because it behaves differently than water when it receives a transfer of energy. In order to raise the temperature of 1.00 g of this stuff by 1.00°C, it takes somewhat less than 1.00 cal of heat. To be exact, at least insofar as we are allowed by the rules of significant figures, it will take 1.00/1.20 = 0.833 cal to raise the temperature of this material by 1.00°C.

Now suppose we have a sample of another material, this time a solid. Let's call it "substance Y." We carve a chunk of it down until we have a piece of mass 1.0000 g, accurate to five significant figures. Again, we use a laboratory balance to determine mass. We transfer 1.0000 cal of energy to substance Y. Suppose the temperature of this solid goes up by 0.80000°C? This material accepts heat energy in a manner different from either liquid water or substance X. It takes a little more than 1.0000 cal of heat to raise the temperature of 1.0000 g of this material by 1.0000°C. Calculating to the allowed number of significant figures, we can determine that it takes 1.0000/0.80000 = 1.2500 cal to raise the temperature of this material by 1.0000°C.

We're onto something here: a special property of matter called the *specific heat*, defined in units of *calories per gram per degree Celsius* (cal/g/°C). Suppose it takes c calories of heat to raise the temperature of exactly 1 g of a substance by exactly 1°C. For pure liquid water, we already know $c = 1$ cal/g/°C, to however many significant figures we want. For substance X above, $c = 0.833$ cal/g/°C (to three significant figures), and for substance Y above, $c = 1.2500$ cal/g/°C (to five significant figures). The value of c is the specific heat for the substance in question.

Alternatively, c can be expressed in *kilocalories per kilogram per degree Celsius* (kcal/kg/°C), and the value for any given substance will be the same. Thus for water, $c = 1$ kcal/kg/°C, to however many significant figures we want. For substance X above, $c = 0.833$ kcal/kg/°C (to three significant figures), and for substance Y above, $c = 1.2500$ kcal/kg/°C (to five significant figures).

THE BRITISH THERMAL UNIT (BTU)

In some applications, a completely different unit of heat is used: the *British thermal unit* (Btu). You've heard this unit mentioned in advertisements for furnaces and air conditioners. If someone talks about Btus literally, in regard to the heating or cooling capacity of a furnace or air conditioner, that's an improper use of the term. They really mean to quote the rate of energy transfer in *Btus per hour* (Btu/h), not the total amount of energy transfer in Btus.

The Btu is defined as the amount of heat energy transfer involved when the temperature of exactly one pound (1 lb) of pure liquid water rises or falls by one degree Fahrenheit (1°F). Does something seem flawed about this definition? If you're uneasy about it, you have a good reason. What is a "pound"? It depends where you are. How much water weighs 1 lb? On the earth's surface, it's approximately 0.454 kg or 454 g. But on Mars it takes about 1.23 kg of liquid water to weigh 1 lb. In a weightless environment, such as on board a space vessel orbiting the earth or coasting through deep space, the definition of Btu is meaningless, because there is no such thing as a "pound" at all.

Despite these flaws, the Btu is still used once in awhile, so you should be acquainted with it. Specific heat is occasionally specified in *Btus per pound per degree Fahrenheit* (Btu/lb/°F). In general this is not the same number, for any given substance, as the specific heat in cal/g/°C.

PROBLEM 1-7

Suppose you have 3.00 g of a certain substance. You transfer 5.0000 cal of energy to it, and the temperature goes up uniformly throughout the sample by 1.1234°C. It does not boil, condense, freeze, or thaw during this process. What is the specific heat of this stuff?

SOLUTION 1-7

Let's find out how much energy is accepted by 1.00 g of the matter in question. We have 3.00 g of the material, and it gets 5.0000 cal, so we can conclude that each gram gets 1/3 of this 5.0000 cal, or 1.6667 cal.

We're told that the temperature rises uniformly throughout the sample. That is, it doesn't heat up more in some places than in other places. It gets hotter to exactly the same extent everywhere. Therefore, 1.00 g of this stuff goes up in temperature by 1.1234°C when 1.6667 cal of energy is transferred to it. How much heat is required to raise the temperature by 1.0000°C? That's the number c we seek, the specific heat. To get c, we must divide 1.6667 cal/g by 1.1234°C. This gives us c = 1.4836 cal/g/°C. Because we are given the mass of the sample to only three significant figures, we must round this off to 1.48 cal/g/°C.

Temperature

Now that we've defined heat, let's be sure we know what we're talking about when we use the term *temperature*. You have a qualitative idea of this. The temperature is generally higher in the summer than in the winter, for example. In quantitative terms, temperature is an expression of the amount of kinetic energy contained in the atoms of a particular sample of matter. In general, for any given substance, as the temperature increases, the atoms and molecules move faster; as the temperature falls, the atoms and molecules move more slowly.

THERMODYNAMIC TEMPERATURE

Thermodynamic temperature is determined according to the rate at which heat energy flows out of a sample of matter into the surrounding environment, or into a sample of matter from the surrounding environment. When energy is allowed to flow from one substance into another in the form of heat, the temperatures "try to equalize." Ultimately, if the energy transfer process is allowed to continue for a long enough time, the temperatures of the two objects will become the same, unless one of the substances is driven away (for example, steam boiling off of a kettle of water).

SPECTRAL TEMPERATURE

Temperature can also be expressed in another way. In order to measure the temperatures of distant stars, planets, and nebulae in outer space, astronomers look at the way they emit electromagnetic (EM) energy in the form of visible light, infrared, ultraviolet, and even radiowaves and x-rays. By examining the intensity of this radiation as a function of the wavelength, astronomers come up with a value for the *spectral temperature* of the distant matter or object.

THE CELSIUS (OR CENTIGRADE) SCALE

Up until now, we've been talking rather loosely about temperature, and have usually expressed it in terms of the Celsius or centigrade scale (°C). This is based on the behavior of water at the surface of the earth, under normal atmospheric pressure, and at sea level.

If you have a sample of ice and you begin to warm it up, it will eventually start to melt as it accepts heat from the environment. The ice, and the liquid water produced as it melts, is assigned a temperature value of 0°C by convention (Fig. 1-6A). As you continue to pump energy into the chunk of ice, more and more of it will melt, and its temperature will stay at 0°C. It won't get any hotter because it is not yet all liquid, and doesn't yet obey the rules for pure liquid water.

Once all the water has become liquid, and as you keep pumping energy into it, its temperature will start to increase (Fig. 1-6B). For awhile, the water will remain liquid, and will get warmer and warmer, obeying the 1 cal/g/°C rule. But eventually, a point will be reached where the water starts to boil, and some of it changes to the gaseous state. The liquid water temperature, and the water vapor that comes immediately off of it, is then assigned a value of 100°C by convention (Fig. 1-6C).

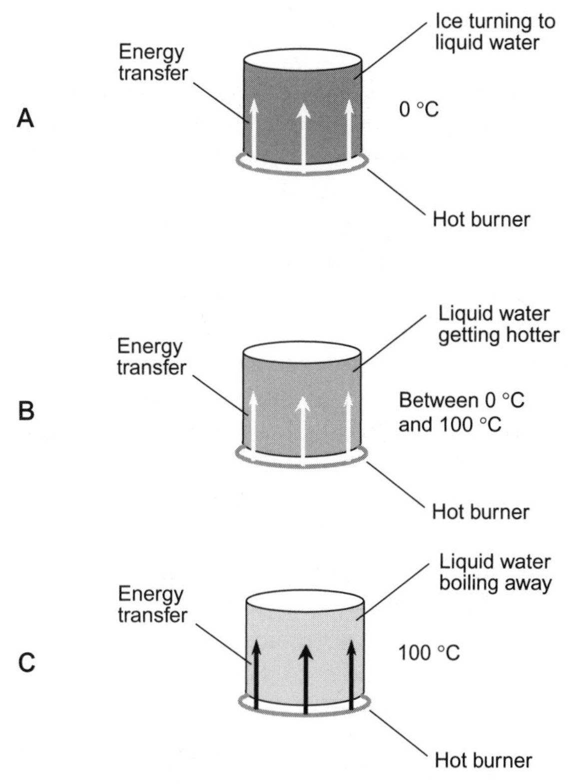

Fig. 1-6. Ice melting into liquid water (A), liquid water warming up without boiling (B), and liquid water starting to boil (C).

Now there are two definitive points—the *freezing point* of water and the *boiling point*—at which there exist two specific numbers for temperature. We can define a scheme to express temperature based on these two points. This is the *Celsius temperature scale*, named after the scientist who supposedly first came up with the idea. Sometimes it is called the *centigrade temperature scale*, because one degree (1 deg or 1°) of temperature in this scale is equal to 1/100 of the difference between the melting temperature of pure water ice at sea level and the boiling temperature of pure liquid water at sea level. The prefix multiplier "centi-" means "1/100," so "centigrade" literally means "graduations of 1/100."

THE KELVIN SCALE

It is possible to freeze water and keep cooling it down, or boil it all into vapor and then keep heating it up. Temperatures can plunge far below 0°C, and can rise far above 100°C. There is an absolute limit to how low the temperature in degrees Celsius can become, but there is no limit on the upper end of the scale. We might take extraordinary efforts to cool a chunk of ice down to see how cold we can make it, but we can never chill it down to a temperature any lower than approximately 273°C below zero (–273°C). This temperature, which represents the absence of all heat, is known as *absolute zero*. An object at absolute zero can't transfer energy to anything else, because it possesses no heat to begin with. There is believed to be no object in our universe that has a temperature of exactly absolute zero, although some atoms in the vast reaches of intergalactic space come close.

Absolute zero is the defining point for the *kelvin temperature scale* (K). A temperature of approximately –273.15°C is equal to 0 K. The kelvin increment is the same size as the Celsius increment. Therefore, 0°C = 273.15 K, and +100°C = 373.15 K. (There is no need to put a plus sign in front of a kelvin temperature figure, because kelvin temperatures are never negative.)

On the high end, it is possible to keep heating matter up indefinitely. Temperatures in the cores of stars rise into the millions of kelvins. No matter what the actual temperature, the difference between the kelvin temperature and the Celsius temperature is always 273.15.

Sometimes, Celsius and kelvin figures can be considered equivalent. When you hear someone say that the core of a star has a temperature of 30,000,000 K, it means the same thing as 30,000,000°C for the purposes of most discussions, because ±273.15 is a negligible difference relative to 30,000,000.

THE RANKINE SCALE

The kelvin scale isn't the only one that exists for defining absolute temperature, although it is by far the most commonly used. Another scale, called the *Rankine temperature scale* (°R), also assigns the value zero to the coldest possible temperature. The difference is that the Rankine degree is the same size as the Fahrenheit degree, which is exactly 5/9 as large as the kelvin or Celsius degree. Conversely, the kelvin or Celsius degree is exactly 9/5, or 1.8 times, the size of the Rankine degree.

A temperature of 50 K is the equivalent of 90°R; a temperature of 360°R is the equivalent of 200 K. To convert any reading in °R to its equivalent in K, multiply by 5/9. Conversely, to convert any reading in K to its equivalent in °R, multiply by 9/5 or 1.8.

The difference between the kelvin and the Rankine scale is significant at extreme readings. If you hear someone say that a star's core has a temperature of 30,000,000°R, they are talking about the equivalent of approximately 16,700,000 K. However, you will not hear people use Rankine temperature numbers very often.

THE FAHRENHEIT SCALE

In much of the English-speaking world, and especially in the United States of America, the *Fahrenheit temperature scale* (°F) is used by lay people. A Fahrenheit degree is the same size as a Rankine degree. However, the scale is situated differently. The melting temperature of pure water ice at sea level is +32°F, and the boiling point of pure liquid water is +212°F. Therefore, a temperature of +32°F corresponds to 0°C, and +212°F corresponds to +100°C. Absolute zero is represented by a reading of approximately –459.67°F.

The most common temperature conversions you will perform involve changing a Fahrenheit reading to Celsius, or vice-versa. Formulas have been developed for this purpose. Let F be the temperature in °F, and let C be the temperature in °C. Then, if you need to convert from °F to °C, use this formula:

$$F = 1.8C + 32$$

If you need to convert a reading from °C to °F, use this formula:

$$C = (5/9)(F - 32)$$

While the constants in the above equations are expressed only to one or two significant figures (1.8, 5/9, and 32), they can be considered mathematically exact for calculation purposes.

Figure 1-7 is a nomograph you can use for approximate temperature conversions in the range from –50°C to +150°C.

When you hear someone say that the temperature at the core of a star is 30,000,000°F, the Rankine reading is about the same, but the Celsius and kelvin readings are about 5/9 as great.

PROBLEM 1-8

What is the Celsius equivalent of a temperature of 72°F?

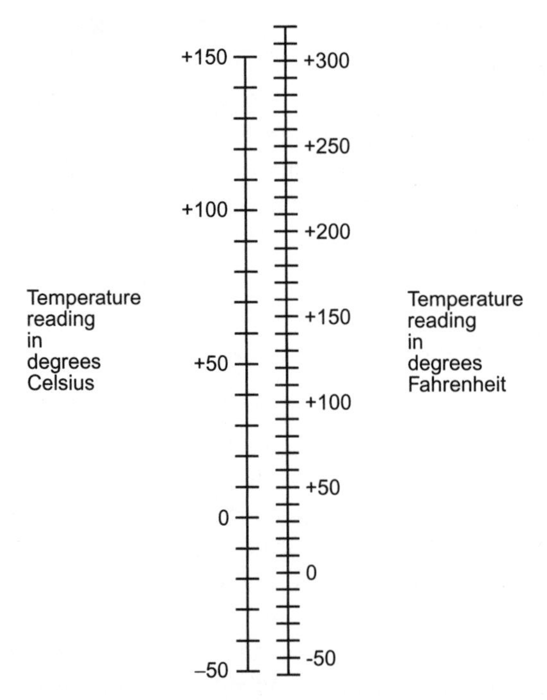

Fig. 1-7. This nomograph can be used for approximate conversions between temperatures in degrees Fahrenheit (°F) and degrees Celsius (°C).

SOLUTION 1-8

To solve this, simply use the above formula for converting Fahrenheit temperatures to Celsius temperatures:

$$C = (5/9)(F - 32)$$

So in this case:

$$C = (5/9)(72 - 32)$$
$$= 5/9 \times 40 = 22.22°C$$

We are justified in carrying this out to only two significant figures, because that is the extent of the accuracy of our input data. So we can conclude that the Celsius equivalent is 22°C.

PROBLEM 1-9

What is the kelvin equivalent of a temperature of 80.0°F?

SOLUTION 1-9

There are two ways to approach this problem. The first is to convert the Fahrenheit reading to Rankine, and then convert this figure to kelvins. The second is to convert the Fahrenheit reading to Celsius, and then convert this figure to kelvins. Let's use the second method, because the Rankine scale is so rarely discussed.

Using the above formula to convert from °F to °C:

$$\begin{aligned} C &= (5/9)(80.0 - 32) \\ &= 5/9 \times 48.0 = 26.67°\text{C} \end{aligned}$$

Let's not round our answer off yet, because we have another calculation to perform. Remember that the difference between readings in °C and K is equal to 273.15. The kelvins figure is the greater of the two. So we must add 273.15 to our Celsius reading. If K represents the temperature in K, then:

$$\begin{aligned} K &= C + 273.15 \\ &= 26.67 + 273.15 \\ &= 299.82 \text{ K} \end{aligned}$$

Now we should round our answer off. We are given our input data to three significant figures, so we can say that the kelvin equivalent is 300 K.

Some Effects of Temperature

Temperature can have an effect on the volume of, or the pressure exerted by, a sample of matter. You are familiar with the fact that most metals expand when they are heated; some expand more than others. This is also true of other forms of matter, including gases and liquids. But for some substances, such as water, the rules aren't as straightforward they are for metals.

TEMPERATURE, VOLUME, AND PRESSURE

A sample of gas, confined to a rigid container, exerts increasing pressure on the walls of the container as the temperature goes up. If the container is flexible (a balloon, for example), the volume of the gas increases as the temperature rises. If you take a container with a certain amount of gas in it and suddenly expand the container without adding more gas, the drop in pressure produces a decrease in the gas temperature. If you have a rigid container with gas in it and then some of the gas is allowed to escape or is pumped out, the drop in pressure will chill the container. This is why a compressed-air canister gets cold when you use it to blow dust out of your computer keyboard.

The volume of liquid water in a kettle, and the pressure it exerts on the kettle walls, don't change when the temperature goes up and down, unless the water freezes. But some liquids, unlike water, expand when they heat up. The element *mercury*, which is a liquid at room temperature even though it is technically a metal, is an example. An old-fashioned *mercury thermometer* works because of this property.

Solids, in general, expand when the temperature rises, and contract when the temperature falls. In many cases you don't notice this expansion and contraction. Does your desk look bigger when the room is 30°C than it does when the room is only 20°C? Of course not. But it is! You don't see the difference because it is microscopic. However, the *bi-metallic strip* in some types of thermostat bends considerably when one of its metals expands or contracts just a tiny bit more than the other. If you hold such a strip near a hot flame, you can watch it curl up or straighten out.

STANDARD TEMPERATURE AND PRESSURE (STP)

In order to set a reference for temperature and pressure, against which measurements can be made and experiments conducted, scientists have defined *standard temperature and pressure* (STP). This is a more-or-less typical state of affairs at sea level on the earth's surface when the air is dry.

The standard temperature is 0°C (32°F), which is the freezing point or melting point of pure liquid water. Standard pressure is the air pressure that will support a column of mercury 0.760 m (just a little less than 30 in) high. This is the proverbial 14.7 pounds per inch squared (14.7 lb/in^2), which translates to approximately 101,000 newtons per meter squared (1.01×10^5 N/m^2).

We don't think of air as having significant mass, but that is because we're immersed in it. When you dive a couple of meters down in a swimming pool, you don't feel a lot of pressure and the water does not feel massive, but if you calculate the huge amount of mass above you, it might scare you out of the water! The density of dry air at STP is approximately 1.29 kg/m^3. A parcel of air measuring 4.00 m high by 4.00 m deep by 4.00 m wide—the size of a large bedroom with a high ceiling—masses 82.6 kg. In the earth's gravitational field, that translates to 182 pounds, the weight of a full-grown man.

THERMAL EXPANSION AND CONTRACTION

Suppose we have a sample of solid material that expands when the temperature rises. This is the usual case, but some solids expand more per degree Celsius than others. The extent to which the height, width, or depth of a solid (its *linear dimension*) changes per degree Celsius is known as the *thermal coefficient of linear expansion.*

For most materials, within a reasonable range of temperatures, the coefficient of linear expansion is constant. That means that if the temperature changes by 2°C, the linear dimension will change twice as much as it would if the temperature changed by 1°C. But there are limits to this. If you heat a metal up to a high enough temperature, it will become soft, and ultimately it will melt, burn, or vaporize. If you cool the mercury in an old-fashioned thermometer down enough, it will freeze. Then the simple length-versus-temperature rule no longer applies.

In general, if s is the difference in linear dimension (in meters) produced by a temperature change of T (in degrees Celsius) for an object whose initial linear dimension (in meters) is d, then the thermal coefficient of linear expansion, symbolized by the lowercase Greek letter alpha (α), is given by this equation:

$$\alpha = s/(dT)$$

When the linear size of a sample increases, consider s to be positive; when the linear size decreases, consider s to be negative. Rising temperatures produce positive values of T; falling temperatures produce negative values of T.

The coefficient of linear expansion is defined in meters per meter per degree Celsius. The meters cancel out in this expression, so the technical unit for the thermal coefficient of linear expansion is a little bit arcane: *per degree Celsius*, symbolized /°C.

PROBLEM 1-10

Imagine a metal rod 10 m long at 20.00°C. Suppose this rod expands by 0.025 m when the temperature rises to 25.00°C. What is the thermal coefficient of linear expansion?

SOLUTION 1-10

The rod increases in length by 0.025 m for a temperature increase of 5.00°C. Therefore, $s = 0.025$, $d = 10$, and $T = 5.00$. Plugging these numbers into the formula above, we get:

$$\alpha = 0.025/(10 \times 5.00)$$
$$= 0.00050/°C = 5.0 \times 10^{-4}/°C$$

We are justified in going to only two significant figures here, because that the limit of the accuracy of the value we are given for s.

PROBLEM 1-11

Suppose $\alpha = 2.50 \times 10^{-4}$/°C for a certain substance. Imagine a cube of this substance whose volume V_1 is 8.000 m^3 at a temperature of 30.0° C. What will be the volume V_2 of the cube if the temperature falls to 20.0°C?

SOLUTION 1-11

It's important to note the word "linear" in the definition of α. This means that the length of each edge of the cube of this substance will change according to the thermal coefficient of linear expansion. The volume changes by a larger factor, because the change in the linear dimension must be cubed.

We can rearrange the above general formula for α so it solves for the change in linear dimension, s, as follows:

$$s = \alpha dT$$

where T is the temperature change (in degrees Celsius) and d is the initial linear dimension (in meters). Because our object is a cube, the initial length, d, of each edge is 2.000 m (the cube root of 8.000, or $8.000^{1/3}$). Because the temperature falls by 10°C, $T = -10.0$. Therefore:

$$s = 2.50 \times 10^{-4} \times (-10.0) \times 2.000$$
$$= -2.50 \times 10^{-3} \times 2.000$$
$$= -5.00 \times 10^{-3} \text{ m} = -0.00500 \text{ m}$$

That means the length of each side of the cube at 20°C is equal to 2.000 – 0.00500 = 1.995 m. The volume of the cube at 20.0°C is therefore $1.995^3 = 7.940149875$ m^3. Because our input data is given to only three significant figures, we must round this off to 7.94 m^3.

Temperature and States of Matter

When matter is heated or cooled, it often does things other than simply expanding or contracting, or exerting increased or decreased pressure. Sometimes it undergoes a *change of state*. This happens when solid ice melts into liquid water, or when water boils into vapor, for example.

THAWING AND FREEZING

Imagine it is late winter in a place such as northern Wisconsin, and the temperature of the water ice on the lake is exactly 0°C. The ice is not safe to skate on, as it was in the middle of the winter, because the ice has become "soft." It is more like slush than ice. It is partly solid and partly liquid. Nevertheless, the temperature of this ice, both the solid and liquid parts, is 0°C.

As the temperature continues to rise, the slush gets softer. It becomes proportionately more liquid water and less solid ice. But its temperature remains uniform at 0°C. Eventually all of the ice melts into liquid. This can take place with astonishing rapidity. You leave for school or work one morning and see the lake nearly "socked in" with slush, and return in the evening to find it almost entirely thawed. Now you can get the canoe out! But you won't want to go swimming. The liquid water near the surface will stay at 0°C until all of the ice is gone.

Now consider what happens in late autumn. The weather, and the water, grows colder. The surface temperature of the water finally drops to 0°C. It begins to freeze. The temperature of this new ice is 0°C. Freezing takes place until the whole lake surface is solid ice. The weather keeps growing colder (a lot colder, if you live in northern Wisconsin). Once the surface is entirely solid ice, the temperature of the ice begins to fall below 0°C, although it remains at 0°C at the boundary just beneath the surface where solid ice meets liquid water (and may be considerably above 0°C deeper down because the lake is fed by subterranean springs). The layer of ice gets thicker. Exposed ice at the surface can get much colder than 0°C. How much colder depends on various factors, such as the severity of the winter and the amount of snow that happens to fall on top of the ice and insulate it against the chill of the air.

The temperature of water does not follow exactly along with the air temperature when heating or cooling takes place in the vicinity of 0°C. Instead, the water temperature follows a curve something like that shown in Fig. 1-8. At A, the air temperature is getting warmer; at B, it is getting colder. The water temperature "stalls" as it thaws or freezes. Many other substances also exhibit this property when they thaw or freeze.

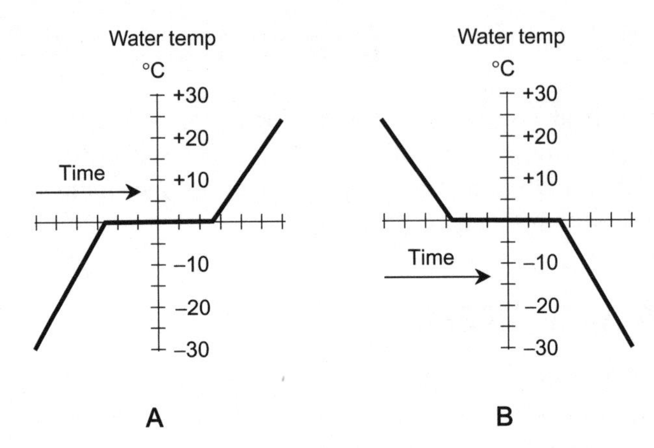

Fig. 1-8. Water as it thaws and freezes. At A, the environmental temperature is getting warmer and the ice is thawing. At B, the environmental temperature is getting colder and the liquid water is freezing.

HEAT OF FUSION

It takes a certain amount of energy to change a sample of solid matter to its liquid state, assuming the matter is of the sort that can exist in either of these two states. (Water, glass, most rocks, and most metals are examples of this kind of matter.) In the case of ice formed from pure water, it takes 80 cal to convert 1 g of ice at 0°C to 1 g of pure liquid water at 0°C. In the reverse scenario, if 1 g of pure liquid water at 0°C freezes completely solid and becomes ice at 0°C, it gives up, or loses, 80 cal of energy. This energy quantity varies for different substances, and is called the *heat of fusion* for the substance.

Heat of fusion is expressed in calories per gram (cal/g). It can also be expressed in kilocalories per kilogram (kcal/kg) and will yield exactly the same numbers as the cal/g figures for all substances. When the substance is something other than water, then the freezing/melting point of that substance must be substituted for 0°C in the discussion.

Heat of fusion is sometimes expressed in calories per mole (cal/mol) rather than in calories per gram. But unless it is specifically stated that the units are intended to be expressed in calories per mole, you should assume that they are expressed in calories per gram.

Suppose the heat of fusion (in calories per gram) is symbolized h_f, the heat added or given up by a sample of matter (in calories) is h, and the mass of the sample (in grams) is m. Then the following relation holds among them:

$$h_f = h/m$$

PROBLEM 1-12

Suppose a certain substance melts and freezes at +400°C. Imagine a block of this material whose mass is 1.535 kg, and it is entirely solid at +400°C. It is subjected to heating, and it melts. Suppose it takes 142,761 cal of energy to melt the substance entirely into liquid at +400°C. What is the heat of fusion for this material?

SOLUTION 1-12

First, we must be sure we have our units in agreement. We are given the mass in kilograms; to convert it to grams, multiply by 1000. Thus $m = 1535$ g. We are given $h = 142{,}761$. Therefore, we can use the above formula directly:

$$h_f = 142{,}761/1535 = 93.00 \text{ cal/g}$$

This is rounded off to four significant figures, because that is the extent of the accuracy of our input data.

BOILING AND CONDENSATION

Suppose a kettle of water is heating up on a stove top. The temperature of the water is exactly +100°C, but it has not yet begun to boil. As heat is continually applied, boiling begins. The water becomes proportionately more and more vapor, and less and less liquid. But the temperature stays at +100°C. Eventually, all the liquid has boiled away, and only water vapor is left. Imagine we have captured all this vapor in an enclosure, and in the process of the water boiling away, all the air has been driven out of the enclosure and replaced by water vapor. The stove burner keeps on heating the water even after all of it has boiled into vapor.

At the moment when the last of the liquid vanishes, the temperature of the water vapor is +100°C. Once all the liquid is gone, the vapor can become hotter than +100°C. The ultimate extent to which the vapor can be heated depends on how powerful the stove burner is, how well insulated the enclosure is, and how much heat the enclosure itself can withstand without breaking, melting, or burning!

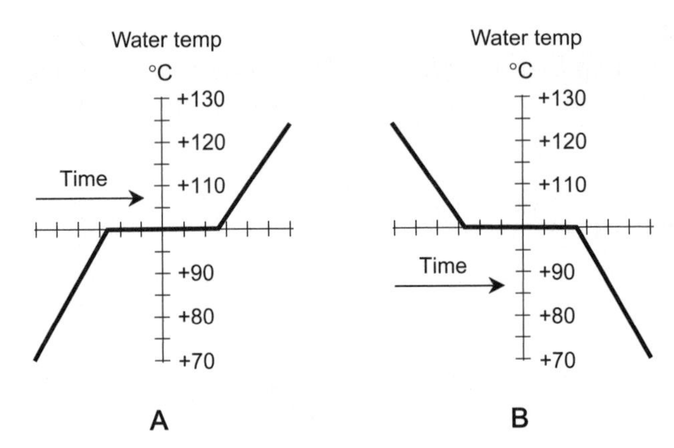

Fig. 1-9. Water as it boils and condenses. At A, the environmental temperature is getting warmer and the liquid water is boiling. At B, the environmental temperature is getting colder and the water vapor is condensing.

Consider now what happens if we take the enclosure, along with the kettle, off of the stove and put it into a refrigerator. The environment, and the water vapor, begins to grow colder. The vapor temperature eventually drops to +100°C. It begins to condense. The temperature of this liquid water is +100°C. Condensation takes place until all the vapor has condensed. We allow a little air into the chamber near the end of this experiment to maintain a reasonable pressure inside. The chamber keeps growing colder; once all the vapor has condensed, the temperature of the liquid begins to fall below +100°C.

As is the case with melting and freezing, the temperature of water does not follow exactly along with the air temperature when heating or cooling takes place near +100°C. Instead, the water temperature follows a curve something like that shown in Fig. 1-9. At A, the air temperature is getting warmer; at B, it is getting colder. The water temperature "stalls" as it boils or condenses. Other substances exhibit this same property when they boil or condense.

HEAT OF VAPORIZATION

It takes a certain amount of energy to change a sample of liquid to its gaseous state, assuming the matter is of a sort that can exist in either of these two states. In the case of pure water, it takes 540 cal to convert 1 g of liquid at +100°C to

1 g of vapor at +100°C. In the reverse scenario, if 1 g of pure water vapor at +100°C condenses completely and becomes liquid water at +100°C, it gives up 540 cal of energy. This quantity varies for different substances, and is called the *heat of vaporization* for the substance.

The heat of vaporization is expressed in the same units as heat of fusion, that is, in calories per gram (cal/g). It can also be expressed in kilocalories per kilogram (kcal/kg) and will yield exactly the same numbers as the cal/g figures for all substances. When the substance is something other than water, then the boiling/condensation point of that substance must be substituted for +100°C.

Heat of vaporization, like heat of fusion, is sometimes expressed in calories per mole (cal/mol) rather than in cal/g. But this is not the usual case.

If the heat of vaporization (in calories per gram) is symbolized h_v, the heat added or given up by a sample of matter (in calories) is h, and the mass of the sample (in grams) is m, then the following formula holds:

$$h_v = h/m$$

PROBLEM 1-13

Suppose a certain substance boils and condenses at +500°C. Imagine a beaker of this material whose mass is 67.5 g, and it is entirely liquid at +500°C. Its heat of vaporization is specified as 845 cal/g. How much heat, in calories and in kilocalories, is required to completely boil away this liquid?

SOLUTION 1-13

Our units are already in agreement: grams for m, and calories per gram for h_v. We must manipulate the above formula so it expresses the heat, h (in calories) in terms of the other given quantities. This can be done by multiplying both sides by m, giving us this formula:

$$h = h_v m$$

Now it is simply a matter of plugging in the numbers:

$$\begin{aligned} h &= 845 \times 67.5 \\ &= 5.70 \times 10^4 \text{ cal} = 57.0 \text{ kcal} \end{aligned}$$

This has been rounded off to three significant figures, the extent of the accuracy of our input data.

Quiz

This is an "open book" quiz. You may refer to the text in this chapter. A good score is 8 correct. Answers are in the back of the book.

1. The heat of fusion is the amount of energy required to
   (a) change a solid into a gas.
   (b) change a gas into a solid.
   (c) dissolve a substance into water.
   (d) None of the above

2. The mass density of chlorine gas is considerably greater than that of air. Because of this, if a lot of chlorine gas is present in a swimming pool enclosure with poor ventilation and poor air circulation, the chlorine gas
   (a) tends to settle near the floor and the pool surface.
   (b) tends to rise to the ceiling.
   (c) does not diffuse into the air at all.
   (d) precipitates out of the air as chlorine bleach.

3. Suppose 2 g of pure water vapor at +100°C condenses completely and becomes liquid water at +100°C. In this process, it
   (a) gives up 540 cal of energy.
   (b) gives up 1080 cal of energy.
   (c) gives up 270 cal of energy.
   (d) does not give up any energy.

4. Imagine a rod made out of some solid substance. Suppose this rod is 10.00 m long at 100°C. Suppose it shrinks to a length of 9.99 m at −100°C. What is the thermal coefficient of linear expansion?
   (a) -2.00×10^{-5}/°C
   (b) 1.00×10^{-5}/°C
   (c) 5.00×10^{-6}/°C
   (d) -2.00×10^{-6}/°C

5. If a substance has a mass of 1 kg and a volume of 1 m^3, what is its mass density in grams per centimeter cubed?
   (a) 1000 g/cm^3
   (b) 10 g/cm^3
   (c) 0.1 g/cm^3
   (d) 0.001 g/cm^3

6. The gravitational pull on Mars is about 37% as strong as the gravitational pull on earth. Suppose a substance has a weight density of 2.00×10^4 N/m^3 on the surface of the earth. What is the weight density of this same substance on the surface of Mars?
   (a) 5.41×10^4 N/m^3
   (b) 7.40×10^3 N/m^3
   (c) 2.00×10^4 N/m^3
   (d) It is impossible to determine this without more information.

7. When you clap two solid chunks of ice together, they don't pass through each other because of
   (a) the fact that solid matter contains no empty space.
   (b) the repulsive electric force produced by the electron shells.
   (c) the high specific gravity of ice.
   (d) the lack of thermal energy in the ice.

8. The gravitational pull on Mars is about 37% as strong as the gravitational pull on earth. Suppose a substance has a particle density of 30 mol/m^3 on the surface of the earth. What is the particle density of this same substance on the surface of Mars?
   (a) 81 mol/m^3
   (b) 11 mol/m^3
   (c) 30 mol/m^3
   (d) It is impossible to determine this without more information.

9. Suppose you buy a household appliance. You save the box, in case you have to return the appliance for service some day. The box is a perfect cube that measures 0.500 m (or 500 mm) on each edge. What is the mass of the air inside the empty box, assuming the air is dry, and is at standard temperature and pressure?
   (a) 161 g
   (b) 323 g
   (c) 645 g
   (d) 1.29 kg

10. Ice floats on the surface of a lake because
   (a) the ice is colder than the water in the lake.
   (b) water ice has a specific gravity lower than 1.
   (c) the ice has a density greater than 1 g/cm^3.
   (d) it is exposed directly to the cold air.

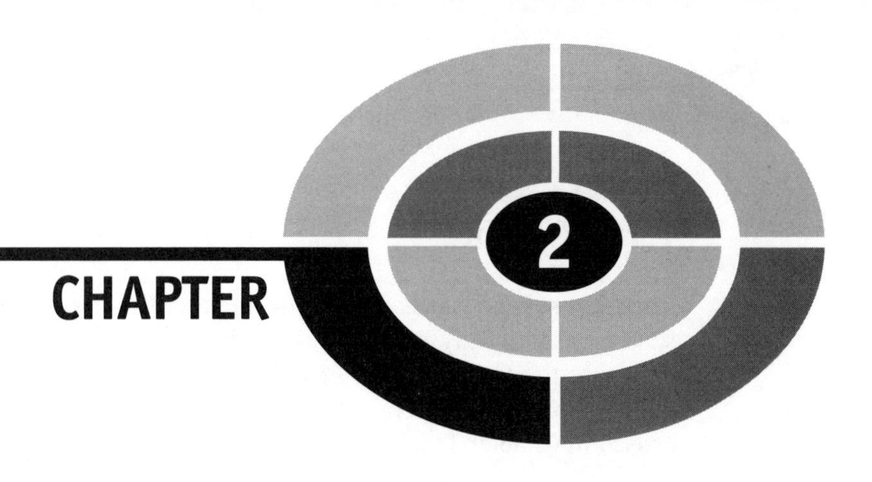

The Atmosphere

We live at the bottom of a dense, swirling, gaseous sea consisting of about 78% nitrogen, 21% oxygen, and 1% other gases and particulate matter including carbon dioxide, argon, neon, water vapor, dust, and pollutants. The sea of air is hundreds of kilometers deep. At the lower levels, it is turbulent and sometimes perilous.

Common Variables

Until recent decades, people had little understanding of the idiosyncrasies of the atmosphere. Before the advent of sophisticated instruments for observing weather phenomena and recording their effects, people who had dwelt for generations in a particular place developed an innate ability to forecast the weather. They used crude instruments, as well as their own senses, to do this. The air pressure, as well as the amount of moisture in the air, seemed to provide good indicators of weather to come. These factors are still used to predict the weather today, and they can be remarkably accurate, especially in regions where weather systems come from the same direction most of the time.

TEMPERATURE

There are several scales for quantifying temperature. In America, the Fahrenheit scale is used by lay people. Scientists, and lay people throughout Europe and Asia, use the Celsius scale. Physicists and chemists often use the kelvin scale, which expresses temperature relative to absolute zero. These scales were discussed in Chapter 1.

BAROMETRIC PRESSURE

You can buy a barometer in a department store for a few dollars and set it according to the local data obtained from a Web site such as the Weather Channel (www.weather.com). The barometer is one of the oldest weather-forecasting instruments. Mariners have used it for centuries. A barometer is especially useful for predicting foul weather, because storm systems are associated with areas of low pressure. As a low-pressure center approaches, the barometer reading falls. If the system is intense (a hurricane, for example) and if you are on a collision course with its center, the barometer reading falls rapidly and dramatically.

In the seventeenth century, an inventor named *Evangelista Torricelli* built a device that demonstrated how the atmosphere exerts pressure on everything. This was the first *mercury barometer.* It was constructed from a glass tube closed off at one end, a small flask, and a few milliliters (cubic centimeters) of mercury. The basic design is shown in Fig. 2-1. You won't see mercury barometers often today, because we know that mercury is toxic, and a mercury barometer needs a large amount of it! But people in Torricelli's time didn't know anything about the dangers of this element.

Torricelli believed that an ocean of air approximately 80 km (50 mi) deep exerted pressure that supported the mercury column and kept it from falling to the bottom of the sealed tube. Today we know that our atmosphere does not abruptly end at an altitude of 80 km; it has no well-defined "surface." It exists in rarefied form at altitudes far greater than 80 km.

DEWPOINT AND FROSTPOINT TEMPERATURE

The air always contains water vapor, but it can only hold a certain amount before it becomes *saturated.* In general, as the temperature of the air rises, its water-vapor–holding capacity increases. The amount of water vapor that the air can

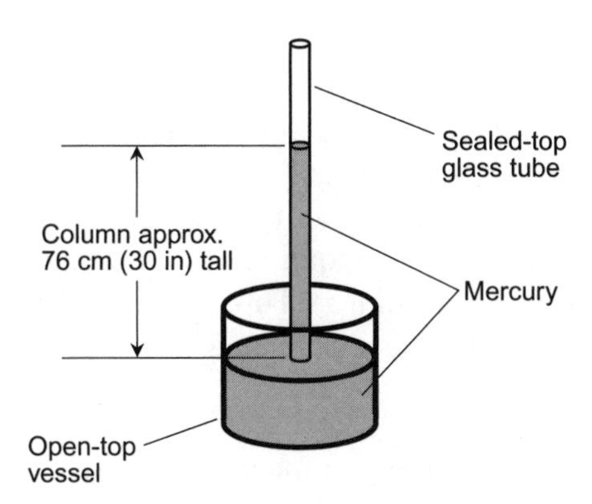

Fig. 2-1. A mercury barometer was made with a glass tube, a flask, and some mercury. Today, you won't see this type of barometer often.

hold roughly doubles with each temperature increase of 10°C (18°F). On a hot summer day, the air can contain much more water vapor than on a cold night in the winter.

Suppose it is a warm day and the air, while containing some water vapor, is far from saturated. Then as evening comes, the air temperature falls, but the amount of water vapor remains nearly constant. At a certain temperature, the air reaches a state in which it is saturated, because cool air has less ability to hold moisture than warm air. When the air, and objects in it, reach this temperature, *condensation* occurs, and *dew* forms. This temperature, called the *dewpoint temperature* (or sometimes simply the *dewpoint*), can be used to quantify the amount of water vapor in the air.

If the temperature must fall below the freezing point of water (0°C or 32°F) before the air becomes saturated, *frost* forms instead of dew. Water vapor is converted directly into ice under these conditions by a process called *deposition,* and the temperature at which frost forms is technically known as the *frostpoint temperature* (or sometimes simply the *frostpoint*). However, the frostpoint is sometimes called the dewpoint, no matter how low the temperature gets.

On nights when there is no wind, dew often forms on grass before it forms on other objects such as cars. This is because the blades of grass "exhale" water vapor by means of *transpiration.* The amount of water vapor within a few centimeters of a grassy lawn, when the air is still, is considerably greater than the amount of water vapor in the air higher up. However, when it is windy, this effect disappears. That is why you are less likely to see dew on grass in the morning after a windy night, as compared with a morning after a calm night.

RELATIVE HUMIDITY

Another way to express the amount of water vapor in the air is to consider the ratio of the amount of water vapor actually in the air (let's call it a), measured in grams of water vapor per kilogram of air (g/kg), to the amount of water vapor in grams per kilogram that would be required for saturation to occur at the same temperature and pressure (let's call it s). This ratio is multiplied by 100 to obtain relative humidity (let's call it h) as a percentage:

$$h = 100a/s$$

Relative humidity is the figure you most often hear in the weather forecasts and reports you get on the radio, on television, or on the Internet.

PROBLEM 2-1

Suppose the relative humidity of the air was 40% on a summer afternoon when the temperature was 20°C. At sunrise the next morning, the air temperature has fallen to 10°C, but the actual amount of water vapor in the air, in grams per kilogram, is the same as it was the previous afternoon. The barometric pressure is also the same as it was the previous afternoon. What is the approximate relative humidity at sunrise?

SOLUTION 2-1

In the afternoon, the ratio a/s was 0.40. This can be seen by "plugging" numbers into the above equation and then manipulating it:

$$\begin{aligned} h &= 100a/s \\ 40 &= 100a/s \\ a/s &= 40/100 = 0.40 \end{aligned}$$

The next morning, the amount of water vapor (in grams per kilogram) required to cause saturation has roughly halved, because the temperature has fallen by 10°C while the barometric pressure has not changed. Therefore, we can substitute $s/2$ for s in the first equation above. Let h^* represent the relative humidity as sunrise. Then:

$$\begin{aligned} h^* &= 100a/(s/2) \\ &= 100(2a/s) \\ &= 100(2 \times 0.40) \\ &= 100 \times 0.80 \\ &= 80 \end{aligned}$$

Because the values of h and h^* are expressed as percentages, the relative humidity at sunrise the next morning is 80%.

SIGNS OF A STORM

Even without a barometer or a *hygrometer* (a device for measuring relative humidity), you can often tell when a storm system is approaching. The wind changes direction. In the tropics of the northern hemisphere, the prevailing east wind turns and blows from a northerly quarter. At northern temperate latitudes, the prevailing west wind shifts counterclockwise to southwest, then south or southeast; this is called a *backing wind.* In the southern hemisphere, the wind also shifts, but in the opposite direction; an east wind becomes southerly, and a west wind becomes northerly.

An approaching storm is commonly preceded by rising temperature and humidity, but not always. A storm coming up the Atlantic coast towards New England will produce an easterly wind off the sea, raising the dewpoint temperature while the barometric pressure falls. The same happens in the Midwest when winds ahead of a storm system blow from the south, bringing moisture from the Gulf of Mexico into the continental interior. However, a hurricane approaching the east coast of Florida from the east may bring dry air in ahead of itself, as backing winds (from easterly to northerly) "pull air down" from the cool North Atlantic, where the air, because of its lower temperature, is capable of holding less moisture. This drives the relative humidity down. It can also produce an unusually clear sky and balmy, pleasant weather, which lulls inexperienced people into complacency. This phenomenon has been called "the calm before the storm." Anyone who has spent a number of years in hurricane-prone regions knows better than to be deceived by this!

OTHER INSTRUMENTS

Meteorologists have sophisticated apparatus for forecasting the weather. Satellites orbit the earth, their cameras scanning the surface for weather patterns. Large systems, which can harbor thundershowers, tornadoes, high winds, heavy snows, or other adverse conditions, have a characteristic signature that the meteorologist recognizes immediately.

As a storm system approaches a weather station, the meteorologist can look at it with a radar set. Large balloons, equipped with instruments to measure temperature, humidity, pressure, wind direction, and wind speed, can be sent aloft to detect atmospheric changes that indicate approaching or developing weather systems. Aircraft pilots can fly their aircraft near (or even into) a storm and see for themselves how bad it is. Meteorologists all over the region, the country, and the world share information, and from this combined effort intricate diagrams are made, showing the locations of air currents, temperature regions, pressure zones, and other parameters.

Atmospheric Circulation

The earth's atmosphere is a complex system that is in continuous motion. Patterns emerge on all scales: planet-wide, hemispheric, regional, and local. The behavior of the atmosphere in any given location also varies with altitude above the surface.

LAYERS OF THE ATMOSPHERE

The lowest layer of the atmosphere, rising from the surface to approximately 18 km (11 mi) altitude, is the *troposphere*. This is where all weather occurs; most of the clouds are found here. In the upper parts of the troposphere, high-speed "rivers of air" travel around the planet. The strongest of these "rivers," called *jet streams,* blow in a generally west–to–east direction between 30 degrees north latitude (30°N) and 60 degrees north latitude (60°N), and between 30 degrees south latitude (30°S) and 60 degrees south latitude (60°S).

Above the troposphere lies the *stratosphere,* extending up to approximately 50 km (30 mi) altitude. Near the upper reaches of this level, ultraviolet (UV) radiation from the sun causes oxygen atoms to group together in triplets (O_3), rather than in pairs (O_2) as is the case nearer the surface. An oxygen triplet is a molecule known as *ozone.* This gas is opaque to most UV rays. Oxygen atoms form a self-regulating mechanism that keeps the surface from receiving too much UV radiation from the sun. Certain gases are produced by industrial processes carried on by humans; these gases rise into the stratosphere and cause the ozone molecules to break apart into their individual atoms. This makes the upper stratosphere more transparent to UV. Some scientists contend that if this process continues, it could have an adverse effect on all life on the planet.

Above the stratosphere lies the *mesosphere,* extending from 50 km (30 mi) to an altitude of 80 km (50 mi). In this layer, UV radiation from the sun causes electrons to be stripped away from atoms of atmospheric gas. The result is that the mesosphere contains a large proportion of charged atoms, or *ions.* This occurs in a layer that engineers and scientists call the *D layer* of the *ionosphere.*

Above the mesosphere lies the highest layer of the atmosphere, known as the *thermosphere.* It extends from 80 km (50 mi) up to more than 600 km (370 mi) altitude. This layer gets its name from the fact that the temperature is extremely high. Ionization takes place at three levels within the thermosphere, called the *E layer,* the *F1 layer,* and the *F2 layer.* Fig. 2-2 is a cross-sectional drawing of the earth's atmosphere, showing the various layers and the ionized regions.

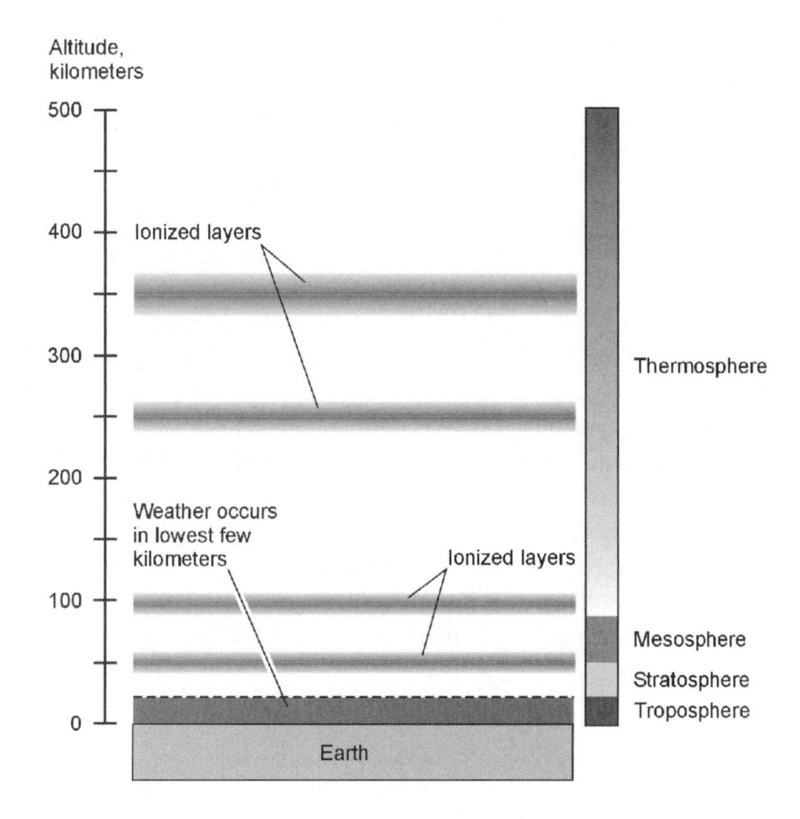

Fig. 2-2. The atmosphere is hundreds of kilometers thick. Weather affecting the surface occurs below approximately 16 km (10 mi) altitude.

HEAT TRANSPORT

The sun warms the earth to a greater extent in some areas than in others, and the air seeks to equalize this imbalance by *convection.* This is what makes the winds and clouds, and is ultimately responsible for all weather.

The equatorial regions receive more heat from the sun than they can radiate back into space. The polar areas are the opposite: They radiate more energy than they get from the sun. The polar/equatorial temperature difference is the result of the astronomical fact that the sun's average angle is more direct at the equator than at the poles. Thus the polar regions have become ice shrouded, and the equatorial zone has not. The different surface characteristics increase the temperature differential still further: Snow and ice reflect much solar energy and absorb little, while vegetation and dark soil reflect little energy and absorb much. If the air did not act to equalize this temperature discrepancy, at least to some extent, the tropics would be boiling hot, and the poles would be incredibly cold.

Fig. 2-3 shows a simplified model that explains how the atmosphere can transport heat on a large scale. The air over the tropics, especially near the equator, is heated by contact with the earth. Therefore, it rises, because warm air always rises. The air over the arctic and antarctic is cold, so it descends, because cold air always descends. The result is that air flows from the poles toward the equator along the planet's surface, and from the equator toward the poles at high altitudes.

According to the model of Fig. 2-3, we should expect to have a prevailing northerly surface wind in the northern hemisphere, and a prevailing southerly surface wind in the southern hemisphere, with no surface winds at the equator or at either pole. This is not what we observe, because Fig. 2-3 is an oversimplification. The air flow around the earth is affected not only by temperature differences, but by geography and the fact that the earth rotates on its axis.

In the actual atmosphere, there are three major convection regions, called *Hadley cells,* in the northern hemisphere, and three "mirror-image" ones in the

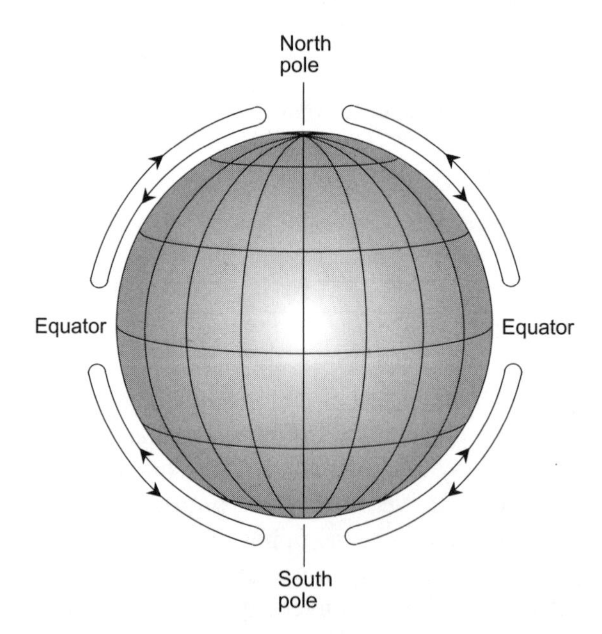

Fig. 2-3. Simplified model showing how heat can be transported from warm zones to cold zones by convection.

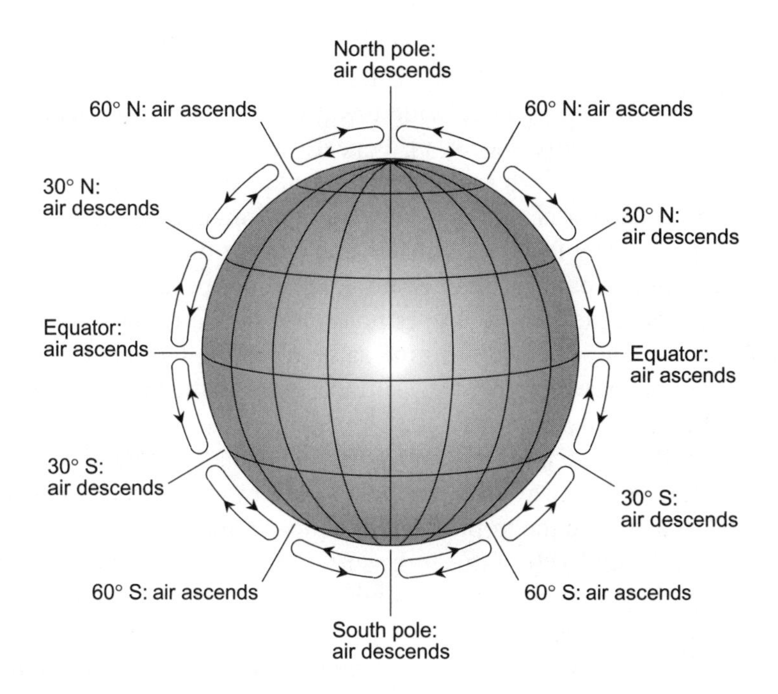

Fig. 2-4. Simplified rendition of convection in the real atmosphere of our planet, showing approximate locations of the Hadley cells.

southern hemisphere. Ascending air is found at or near the equator and at about 60°N and 60°S. These areas are *semipermanent low pressure systems,* and the weather in and around them is characterized by abundant cloudiness and rainfall. Descending air is found at or near the poles and at about 30°N and 30°S. The descending air is associated with *semipermanent high-pressure systems,* with predominantly fair weather. This convection pattern, neglecting the effects of the earth's rotation, is shown in Fig. 2-4. But even this is an oversimplification.

EASTERLIES AND WESTERLIES

Between the equator and about 30°N, the air at the surface flows generally toward the equator, from a zone of relatively high pressure to a zone of relatively low pressure. The same thing happens between the equator and about 30°S. As the air reaches progressively lower latitudes (that is, it gets closer to the equator), the *tangential speed* of the earth's surface, resulting from the rotation of the planet, increases. Therefore, air flowing toward the equator is deflected toward the west; the earth literally speeds up beneath it. We find prevailing easterly winds in these regions, known as the *torrid zones* or the *tropics,* for this reason.

Between 30°N and 60°N, and also between 30°S and 60°S, the air at the surface tends to flow away from the equator and toward the pole, from the semipermanent 30° high-pressure zone toward the semipermanent 60° low-pressure zone. As the air near the surface moves poleward, the earth slows down underneath it. Thus, the air is deflected in the same direction the planet rotates. For this reason, we find prevailing westerly surface winds in the mid-latitude regions, which are called the *temperate zones.*

Between 60°N and the north pole, and also between 60°S and the south pole, the air flow is generally away from the poles and toward the equator. The effect here is the same as it is in the tropics, but of course at much colder temperatures. The earth speeds up under the air near the surface as it flows away from the pole. There are prevailing easterly surface winds in the arctic and antarctic regions, which are called the *frigid zones.*

The prevailing winds of the torrid, temperate, and frigid zones sometimes shift because of changes in the locations of the semipermanent high and low pressure belts. In general, however, the torrid zones have easterly winds called the *trade winds,* the temperate zones have westerly winds called the *prevailing westerlies,* and the frigid zones have easterly winds called the *polar easterlies.* The earth's atmosphere is thus broken up into six major belts or zones that girdle the planet.

SEMIPERMANENT PRESSURE REGIONS

Near the 30° and 60° latitude lines, both in the northern hemisphere and in the southern hemisphere, the surface winds converge from, or diverge in, opposing directions. This gives the air an impetus to spin clockwise at 30°N and 60°S, counterclockwise at 60°N and at 30°S. The air also spins around the poles, clockwise in the arctic and counterclockwise in the Antarctic. The eddies near the 60th parallels tend to pull surface air inward, and the eddies near the 30th parallels and the poles tend to push surface air outward.

Because of the tendency of the air to rotate near the 30th and 60th parallels and the poles, we find persistent *cyclonic* (inward-spiraling) and *anticyclonic* (outward-spiraling) atmospheric zones. The cyclonic zones are semipermanent low-pressure systems, and the anticyclonic zones are semipermanent high-pressure systems.

Over the oceans, semipermanent low-pressure regions are found near the Aleutian islands in the North Pacific, Iceland in the North Atlantic, and just off the coast of Antarctica in the South Pacific, South Atlantic, and Indian Oceans. Semipermanent high-pressure systems are found near Hawaii in the North Pacific, Bermuda in the North Atlantic, Tahiti in the South Pacific, between South America and Africa in the South Atlantic, and between Australia and Africa in the Indian Ocean.

Over land masses, the semipermanent high and low pressure systems are not as stable as they are over the oceans. This is because temperature variations are greater over land masses, and temperature changes affect atmospheric circulation. Nevertheless, certain land regions, notably deserts, are usually dominated by semipermanent high pressure. Sometimes, a high- or low-pressure system can stall over a land mass for a considerable period of time. This brings droughts or wet spells to such regions.

A semipermanent low-pressure region, called the *intertropical convergence zone* (ITCZ), exists near the equator in a sinuous band that completely circles the globe. Semipermanent highs are found near the poles. The ITCZ, in particular, is associated with major weather events, particularly hurricanes, that develop over warm ocean waters at certain times of the year.

PROBLEM 2-2

Are Hadley cells, along with semipermanent highs and lows and prevailing winds, unique to the earth?

SOLUTION 2-2

No. Similar patterns are observed on other planets that have atmospheres. On Jupiter and Saturn, each hemisphere has multiple cells. The atmosphere of Jupiter is an especially good example, because the belts and zones of prevailing winds show up as light and dark cloud bands. These bands can be seen through a good telescope on a clear night.

Weather Systems

Semipermanent oceanic highs were well known to mariners who sailed between Europe and the New World or the Orient. Seamen, becalmed in the hot, fair weather near 30°N or 30°S, sometimes remained motionless for days. Rations ran short. The men got hungry and ate the food intended for the horses they had brought with them, and threw the horses overboard to drown. Sometimes the hapless sailors even ate the horses themselves! This, according to some legends, is how the regions became known as the *horse latitudes.*

SPINOFFS

A set of intense, persistent low-pressure systems dominate the climate over the oceans near 60°N and 60°S. These great lows spawn low-pressure systems called *frontal cyclones* that sweep from west to east across the continents. One such low-pressure system, which is largely responsible for much of the changeable weather in the United States and Canada, is called the *Aleutian low* because it is centered just off the southern coast of Alaska. In winter, the Aleutian low generates frontal cyclones that drench the west coast of North America with one storm after another. Some of these cyclones rival hurricanes for ferocity.

Another well-known semipermanent low exists near Greenland and Iceland. This low, like its Aleutian cousin, produces smaller, frontal cyclones that move toward the east and southeast. The *Greenland–Iceland low* intensifies during the fall and winter, so the climate of northern Europe is not unlike that of the Pacific Northwest during these months. Some winter gales in this region, like their counterparts in the Pacific Northwest, produce tidal floods and damaging winds.

The Aleutian and Icelandic lows, although well known, cannot rival the chain of semipermanent low-pressure systems that dominate the southern hemisphere between approximately 40°S and 70°S. The *Tierra del Fuego* region, at the southern tip of South America, is regularly blasted by gales. The stormy *Strait of Magellan,* named after the European explorer who sailed through the region on his quest to circumnavigate the world, is respected by sailors to this day. The westerly winds of the southern hemisphere blow around the planet almost unobstructed by land masses between 40°S and 50°S, earning this region the nickname the *roaring forties.*

JET STREAMS

The general circulation around the semipermanent lows is counterclockwise in the northern hemisphere and clockwise in the southern hemisphere. The air circulation around the semipermanent highs is just the reverse: clockwise in the northern hemisphere and counterclockwise in the southern hemisphere. This produces a prevailing west–to–east wind in the temperate latitudes, with which most of us are familiar. The lows dominate the weather much of the time between 45°and 60° north or south latitude; the highs control the weather much of the time between 30° and 45°. The boundaries between the influence of low and high pressure are sharp, and they constantly fluctuate, moving closer to the equator in the hemispheric winters and closer to the poles in the hemispheric summers.

During the Pacific battles of World War II, aviators often observed that their air speeds and ground speeds were different, sometimes by as much as 300 km/h (about 200 mi/h). These aviators had discovered a jet stream. Today, passenger airliners flying in the middle latitudes are sometimes ahead of schedule or behind schedule because of the mid-latitude jet stream, which flows from west to east and marks the boundary between the semipermanent pressure systems.

The mid-latitude jet stream is recognized as a major factor that contributes to weather in the United States, Canada, Europe, and Asia. North of the jet stream in the northern hemisphere, polar air masses dominate. South of the jet stream, tropical air prevails. The situation is reversed in the southern hemisphere. A semipermanent front encircles the whole planet between 30° and 60°N. A similar front exists between 30° and 60°S. The mid-latitude jet streams in both hemispheres flow at an altitude of several kilometers near the top of the troposphere, and can vary greatly in width and speed.

POLAR FRONTS

In either hemisphere, the semipermanent front, called the *polar front,* makes unending attempts to invade the tropics, pushing the jet streams toward the equator. The tropical air fights back, limiting the progress of the polar front and occasionally causing the jet stream to move toward the pole. The polar front changes position almost every day. The polar air mass expands during the winter months and shrinks in the summer. The mid-latitude jet stream thus lies closer to the equator in the winter than in the summer.

Much of the United States and all of Canada are in the grip of the polar air mass during the winter. Most of Europe and much of Asia endure the same fate. In the southern hemisphere, much less land area is invaded by the polar front from Antarctica, but this is because most of the inhabited lands in that hemisphere lie far from the pole.

Sometimes certain areas are spared the domination of the polar air mass for weeks or even months during the winter season. At other times, the polar front moves unusually close to the equator in a particular part of the world, and a cold wave occurs. An event of this type occurred in December of 1983, during one of the most severe early winters on record. The midwestern United States had readings that stayed well below zero Celsius for days in a row, in some locations reaching –40°C (–40°F). In Florida and Texas, oranges and grapefruit froze. Even in Miami, frost was observed.

The polar front does not invade the tropics everywhere at the same time. The United States may suffer while Europe stays comfortable. While Anchorage, Alaska, experiences record high temperatures, New York City might experience record lows. On the morning of December 26, 1983, the temperature in Miami, Florida, was colder than it was in Moscow, Russia. The jet stream dipped toward the south over the United States, forming what meteorologists call a *trough.*

Air masses also do battle in the summer, but then the tropical air goes on the offensive. The cold polar front retreats, reverses direction, and becomes a warm *tropical front.* The jet stream is generally more stable in summer than in winter. The southern United States, and much of southern Europe, settles into a weather pattern that stays fairly constant from early June into September. At times, the tropical air pushes the jet stream far toward the pole, and a large part of North America or Europe stays under its influence for weeks. The jet stream moves along with the tropical front over the land, and forms what meteorologists call a *ridge.* The barometric pressure rises and remains high. The weather under these conditions can become extremely hot and dry. This kind of situation dominated the summer weather during 1983, pushing temperatures to all-time record highs and drying out farm crops in the Great Plains of North America.

MEANDERS IN THE JET STREAM

The jet stream flows in an irregular path, not straight around the world at one latitude. An experiment can be conducted to demonstrate this effect on a small scale, using a rotating pan, ice, water, and a heating element. The pan is filled with water of a moderate temperature. A container of ice water is placed at the center of the pan, and the heating element is placed under the periphery of the pan, all the way around (Fig. 2-5). This sets up a "battle" between cold water and hot water. Sawdust, or some other fine substance that will float, is sprinkled on the surface of the water, facilitating observation of the water circulation. The pan is then spun around.

Although the pan is circular, and not spherical like the earth, the forces imposed on the water are similar to the forces that occur in the atmosphere of our planet. The center of the pan does not move but simply rotates, as does the north or south pole of the earth. The outer edge of the pan has considerable tangential speed, as does the equator. The water in the pan develops a miniature

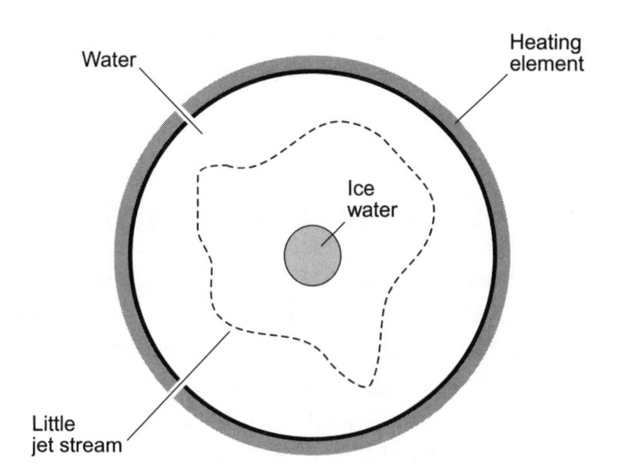

Fig. 2-5. An artificial jet stream can be produced in a rotating pan of water.

"jet stream" somewhere between the center and the periphery, and this stream flows in the direction of rotation, as can be seen by watching the sawdust. If the pan is rotated fast enough, the stream develops irregularities that resemble the meanders of the jet stream in the real atmosphere.

If the temperature of the heating element is increased, simulating the conditions of summer, the stream becomes smaller, moving in toward the center of the pan. If the heat is turned down and extra ice is placed in the container at the center, the stream moves toward the periphery of the pan, just as the real jet stream moves toward the equator in the winter.

The jet stream flows in an irregular way because of inherent instability of the system. A bend in the jet stream is called *cyclonic* if it turns toward the left (counterclockwise) in the northern hemisphere or toward the right (clockwise) in the southern hemisphere. A bend is termed *anticyclonic* if it turns toward the right in the northern hemisphere, or toward the left in the southern hemisphere. When the polar air mass pushes toward the equator, a *cyclonic bend* is produced in the jet stream; when the tropical air mass advances on the pole, an *anticyclonic bend* is produced.

Cold low-pressure systems on the polar side of a cyclonic trough in the jet stream tend to be stable and persistent. The same is true of warm high-pressure systems that exist on the equatorial side of an anticyclonic ridge. Troughs and ridges can form over the oceans or over the continents. Persistent ridges tend to develop over the oceans during winter and over land in the summer. Troughs tend to form over the continents in the winter. Continental ridges bring warm, dry, fair weather, and sometimes they produce heat waves. Troughs generate stormy weather.

WHEN AIR IS SQUEEZED AND STRETCHED

The jet stream varies in width and speed, as well as in direction. These variations cause air to be pushed together, or compressed, in some places causing an increase in pressure. In other places, the air is pulled apart, and the pressure goes down. Either of these situations can result from changes in the width of the jet stream, its speed, or both. When air is pushed together, it is said to be *converging,* and when it is pulled apart, it is said to be *diverging.*

Fig. 2-6 illustrates convergence and divergence. *Directional convergence* and *directional divergence* are shown at A and B, respectively. *Speed convergence* and *speed divergence* are shown at C and D, respectively.

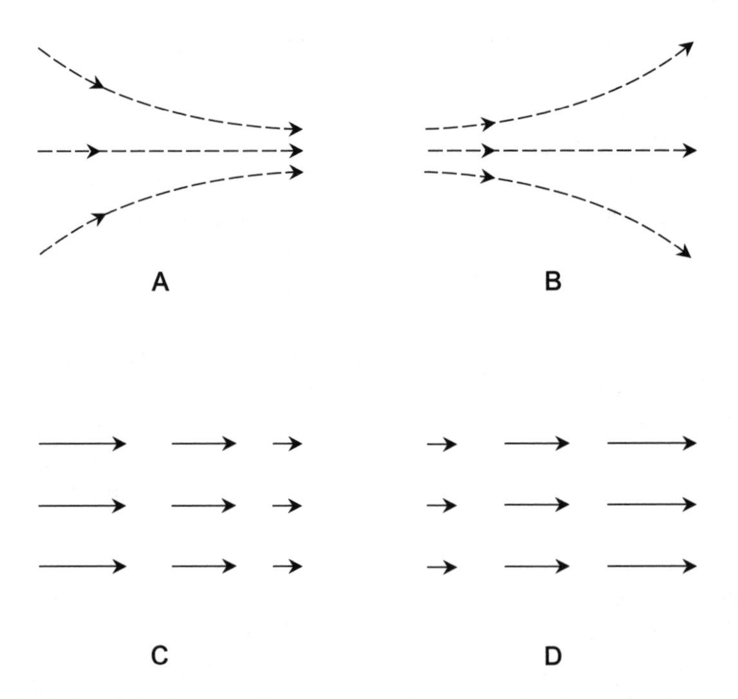

Fig. 2-6. At A, directional convergence of air. At B, directional divergence. At C, speed convergence. At D, speed divergence.

Directional effects take place only at bends in the jet stream. But speed effects can occur whether there is a bend or not. Divergence gives rise to low-pressure systems, because divergence creates a drop in pressure. If the air in a particular part of the jet stream diverges to a great enough extent, clouds form and the weather becomes foul. The system is carried along, from west to east, by the jet stream. Warm tropical air moves toward the pole ahead of the system, and cold polar air flows in behind the low-pressure center. The result is a *frontal cyclone,* also known as a *low.* This type of system can form anywhere along the jet stream, provided there is enough divergence of the air.

The most intense frontal cyclones are generated when there is not only strong divergence in the air, but also a cyclonic bend in the jet stream. If the jet stream is turning in a cyclonic direction (toward the left in the northern hemisphere, or toward the right in the southern) when a frontal cyclone forms, the circulation is given an extra push. If an existing frontal cyclone encounters a cyclonic bend in the jet stream, intensification is likely.

Frontal cyclones are always accompanied by clouds. As a low-pressure system approaches, the first sign is an increase in high-altitude, thin clouds. The clouds thicken and become lower, until the sky is a dull overcast and rain or snow falls. If the center of a frontal cyclone in the northern hemisphere passes to the south, the weather becomes chillier, and the winds shift from south to east, then to the north and northwest. If the center of the system passes to the north, the rain or snow may abate, and the temperature usually rises. The wind shifts to the southwest. A day or two later, the sky becomes cloudy again. In the summer, fair-weather clouds give way to towering *cumulonimbus,* and heavy thunderstorms are common. In the winter, freezing rain or snow falls, driven by strong westerly or northwesterly winds.

TROPICAL SYSTEMS

The mid-latitude jet stream, marking the battle front between the tropical and the polar air masses, is responsible for changeable weather in the temperate zone. In the tropics, far from the influence of the mid-latitude jet stream, the prevailing winds are from east to west. The weather in the tropics is relatively stable from day to day; the trade winds blow at a nearly constant speed of 15 to 30 km/h (approximately 10 to 20 mi/h), and fair-weather cumulus clouds dot the sky. Near the equator, these clouds tower higher and higher as morning progresses into afternoon, producing daily rains in the jungle over land, and squally downpours of the ITCZ over the oceans.

The weather in the mid-tropical latitudes (about 10°N to 25°N and 10°S to 25°S) is not always benign, as anyone who has lived there knows. The tropics are plagued by their own breed of cyclone. The equator receives far more energy from the sun than it can radiate into space. Sometimes the normal atmospheric circulation is not enough to maintain equilibrium. It is then that the ITCZ makes its effects felt in the form of a *tropical cyclone.* The system begins as a trough, with cyclonic deflection of the trade winds. Clouds and rain accompany this so-called *easterly wave,* and the system moves from east to west. If intensification occurs, a closed, rotating low-pressure system, called a *tropical depression,* forms, and the precipitation becomes heavier. The circulation may then become concentrated, producing the high winds and torrential rains of a *tropical storm* or *hurricane.*

Tropical weather systems help to transfer heat from the equator toward the poles. The systems also provide much-needed rainfall in the normally dry areas of Texas and Mexico. But a major hurricane can inflict destruction because of the high winds, which can reach sustained speeds of around 320 km/h (200 mi/h) in extreme cases. More severe damage is caused by the rough and high seas that are generated by the winds and the low barometric pressure.

POLAR SYSTEMS

The north pole is covered by an ocean that is almost completely surrounded by land masses. The south pole, in contrast, lies near the middle of a continent. The polar regions are dominated by high-pressure systems. These air masses produce prevailing easterly winds. At the north pole, the high-pressure system sometimes weakens, and the prevailing winds slacken and may even reverse direction.

Both polar regions have a dry climate. On the average, they receive less than 25 cm (10 in) of total melted precipitation annually. This qualifies them as true deserts. As the distance from either pole increases, the effects of the semipermanent low-pressure belts near 60°N and 60°S become increasingly apparent. Cloudy skies, frequent fog, and abundant precipitation become more common. The winds increase in speed and become more variable. In summer, the low-pressure belt in either hemisphere moves slightly closer to the pole; in winter, it moves toward the equator.

The weather in the arctic and antarctic, although dominated by semipermanent high pressure, is not gentle. Ask anyone who has lived in northern Alaska, or who has worked at one of the scientific installations in Antarctica. Blizzard conditions with high winds occur often. In the antarctic, the wind speeds rival those in major hurricanes. Small, intense low-pressure systems, sometimes called *arctic hurricanes,* have been observed near the periphery of the high pressure region surrounding the north pole.

PROBLEM 2-3

Why are colder temperatures observed in the antarctic winter, as compared with the arctic winter?

SOLUTION 2-3

The antarctic region is dominated by a land mass, which cools down more rapidly, and to a greater extent, than water in the absence of solar irradiation. The arctic is dominated by an ocean with outlets to other oceans, and this serves to keep the temperature higher and more constant than it would be if the region were covered by land.

Weather Maps

Because of the endless movement of weather systems throughout the world, especially in the temperate latitudes, conditions vary greatly from place to place. Before wire or radio communications were available, a person living in New York City could only guess at the current weather in Philadelphia or Boston. The conditions at any given place can change within minutes.

After the telegraph was invented, it was possible to get weather information almost instantly. But storms regularly tore the wires down. Then came the invention of the "wireless telegraph." Only a few of the more violent frontal cyclones managed to rip down radio antennas and cut off communications, but the sun occasionally produced (and still produces) ionospheric and geomagnetic disturbances that degraded or ruined radio-wave–propagation conditions.

THE STATION MODEL

Today, every city has a weather station that is linked by wire, fiberoptic cable, terrestrial radio, and satellite systems to a central office. The meteorologist in Rochester, Minnesota, can find out in a matter of seconds what happening in Eureka, California. The data from a particular place includes temperature, dewpoint, extent and type of cloud cover, type of precipitation (if any), wind direction, wind speed, and barometric pressure. All of this information is fed into a computer, and the computer transforms it into a detailed graphical representation of the weather over the whole country. The conditions at each weather station are indicated on the weather map by a set of symbols known as a *station model.*

ISOBARS

Equal-pressure lines, called *isobars,* are drawn on weather maps to represent various pressure levels in units called *millibars* (mb). Isobars are typically drawn representing pressures at intervals of 4 mb, equivalent to about 3 mm (0.12 in) of mercury. If P_{mb} is the pressure in millibars, P_{mm} is the pressure in millimeters of mercury, and P_{in} is the pressure in inches of mercury, then you can use the following formulas to convert among them:

$$P_{mb} = 1.333\ P_{mm} = 33.86\ P_{in}$$
$$P_{mm} = 0.750\ P_{mb} = 25.40\ P_{in}$$
$$P_{in} = 0.02953\ P_{mb} = 0.03937\ P_{mm}$$

The isobars on a weather map show the locations of the cyclones and anticyclones. In the vicinity of high- and low-pressure systems, the isobars are curved. If the system is intense, several closed isobars exist around the center. The wind blows nearly parallel to the isobars, especially when the isobars are close together. When isobars are close together, it shows that the *pressure gradient* is steep. High winds tend to occur in regions indicated by closely spaced isobars on the weather map.

SYMBOLS IN GENERAL

The temperature data on a weather map gives an indication of the locations of *fronts,* where air masses having different characteristics come together. Most severe weather occurs near fronts. The isobars are “kinked” along the line of an intense or fast-moving front. Stormy weather tends to occur near these zones. Weather fronts are plotted as lines with bumps or barbs that indicate the type of front and the direction of movement.

There are four major types of fronts. In a *cold front,* a cold air mass pushes its way into warmer air. In a *warm front,* a warm air mass pushes its way into

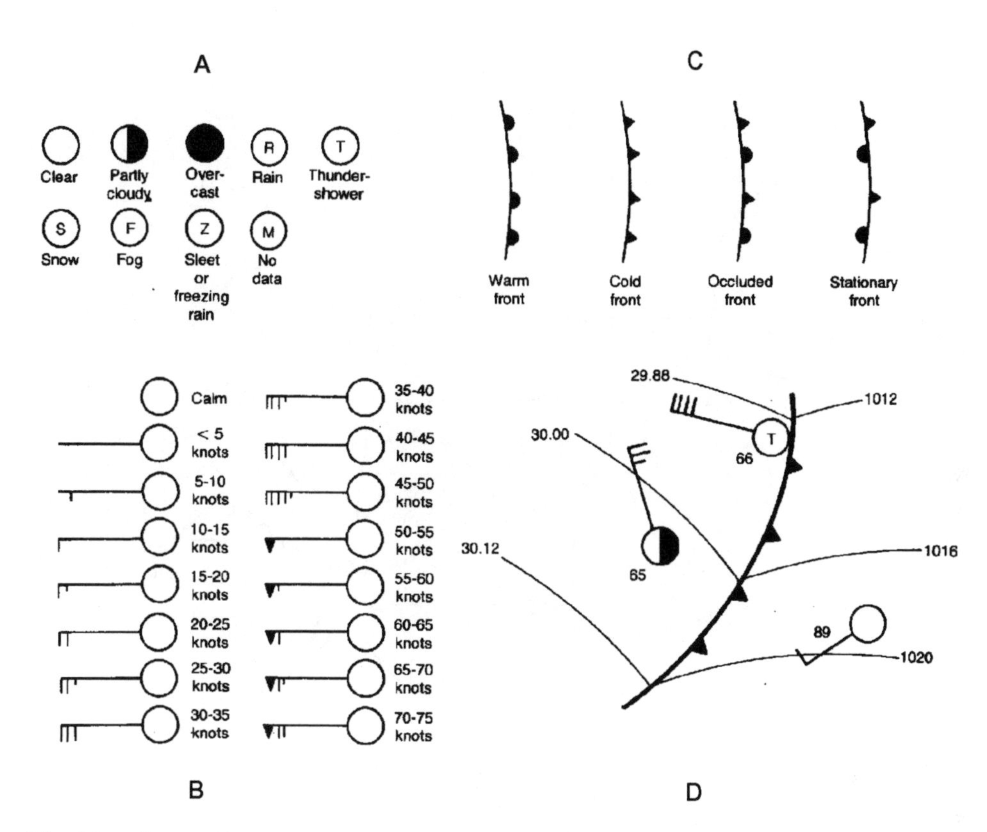

Fig. 2-7. Sky condition symbols (A), wind speed symbols (B), frontal symbols (C), and a cold front as it might appear on a local weather map (D).

cooler air. In a *stationary front,* a warm air mass and a cold air mass meet in a boundary that moves slowly or not at all. When a cold front catches up to a warm front ahead of it, the result is called an *occluded front.*

The common weather map symbols are shown in Fig. 2-7, and a hypothetical weather map, similar to the type you might see in a major daily newspaper, is shown in Fig. 2-8. In the situation of Fig. 2-8, a strong midwinter low-pressure system is sweeping across the central United States, and an intense cold front is pushing eastward across Texas.

The weather maps used by meteorologists are more detailed than the one shown in Fig. 2-8. Computer-generated maps show such things as lines of equal temperature (*isotherms*) or weather conditions at various altitudes, but the simple

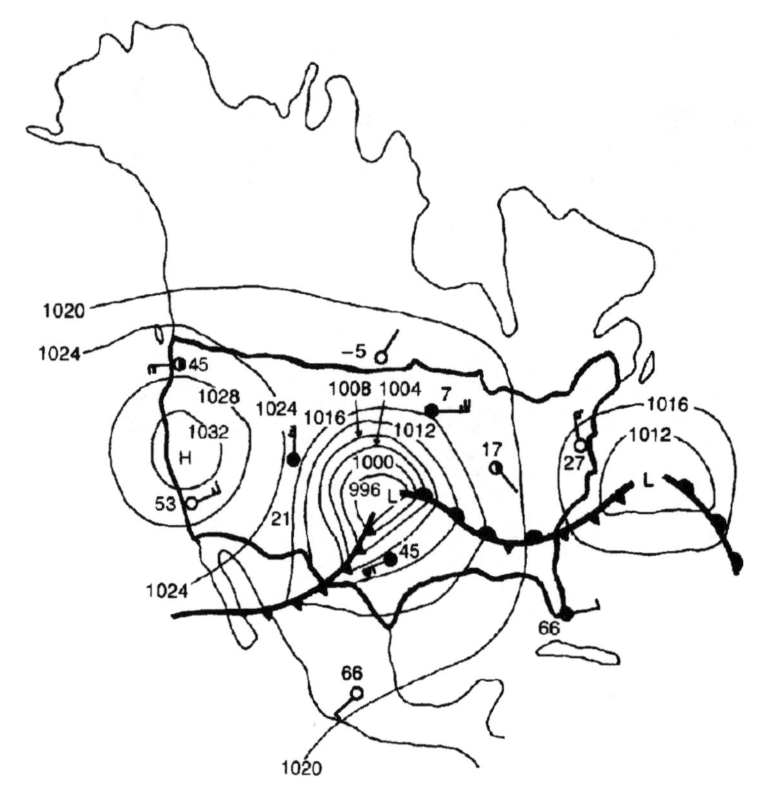

Fig. 2-8. A simplified weather map, showing a hypothetical low-pressure system over the central United States.

map found in the daily newspaper is sufficient for you to get a good idea of what is happening throughout the continent. You can tell with reasonable accuracy where the major weather systems are. If you know the position of the jet stream, you can forecast where the storms will pass.

PROBLEM 2-4

In the situation shown by Fig. 2-8, where would the strongest widespread winds likely be found?

SOLUTION 2-4

The most intense winds are likely to occur where two things happen simultaneously: (1) the isobars on the weather map are close together, and (2) the wind direction is the same as the direction in which the weather system giving rise to that wind is moving. In Fig. 2-8, this is near, and to the south of, the center of the low-pressure system over the middle part of the United States.

PROBLEM 2-5

In the situation shown by Fig. 2-8, where would the strongest localized winds likely be found?

SOLUTION 2-5

This question is more difficult to answer than the previous one. Localized winds are affected by mountains, canyons, the presence of tall buildings, and storms associated with weather fronts. Cold fronts are often associated with heavy thunderstorms, which can occur at any time of year in the extreme southern United States. It is even possible that there is a tornado somewhere near the cold front in the situation shown by Fig. 2-8, in which case the highest localized winds would be found there.

Wind Speed

Normally, the wind does not blow hard enough to have too much effect on things. Occasionally, there is an exceptionally windy day or a brief rush of strong wind as a thundershower approaches and passes. Once in a while, the wind becomes strong enough to cause massive damage to natural and human-made things.

UNITS OF SPEED

Wind speed can be specified in *meters per second* (m/s), *kilometers per hour* (km/h), *statute miles per hour* (mi/h), or *nautical miles per hour,* also known as *knots* (kt). The most common unit used by weather forecasters and other professionals is the knot, which is equivalent to approximately 1.852 km/h or 1.151 mi/h. The most common unit used by news and weather broadcasters in the United States is the statute mile per hour, which is equivalent to approximately 1.609 km/h or 0.8690 kt.

THE BEAUFORT SCALE

Early in the nineteenth century, an admiral in the British Navy, *Sir Francis Beaufort,* noticed that winds of various speeds produced consistent and visible effects on land and at sea. Not everyone has a wind-speed measurement instru-

Table 2-1. Beaufort scale for winds on land. Wind speeds are sustained (based on a 1-minute average.)

Beaufort number	Wind speed in knots	General observed effects
0	0	Calm; smoke rises vertically.
1	1–3	Smoke shows direction of wind, but weather vanes do not.
2	4–6	Wind can be felt on face. Leaves move slightly.
3	7–10	Leaves are in continual motion. Small flags unfurl.
4	11–16	Dust is raised. Papers blow around. Small tree branches move.
5	17–21	Medium-sized tree branches move. Small trees sway.
6	22–27	Large tree limbs are in constant motion. Utility wires whistle.
7	28–33	Whole trees are in constant motion. Umbrellas turn inside-out.
8	34–40	Gale. Twigs and leaves break off trees. Walking into wind is difficult.
9	41–47	Strong gale. Branches break off trees.
10	48–55	Whole gale. Trees are heavily damaged. Minor structural damage occurs.
11	56–63	Storm. Numerous trees blow down. Considerable structural damage occurs.
12	64 or more	Hurricane. Widespread destruction occurs.

Table 2-2. Beaufort scale for winds over large bodies of water. Wind speeds are sustained (based on a 1-minute average).

Beaufort number	Wind speed in knots	General observed effects
0	0	Surface is glassy, but swells may exist.
1	1-3	Small ripples form.
2	4-6	Small waves form.
3	7-10	Moderate-sized waves form. Some waves break.
4	11-16	Moderate-sized waves break consistently.
5	17-21	Moderate-sized and large waves occur, and some airborne spray is observed.
6	22-27	Large waves occur with whitecaps and airborne spray.
7	28-33	Large waves produce foam, and there is considerable airborne spray.
8	34-40	Gale. Large waves occur with foam, streaking on surface, and airborne spray.
9	41-47	Strong gale. High waves occur with dense foam streaks. Air is dense with spray.
10	48-55	Whole gale. High waves occur with dangerous crests. Visibility deteriorates.
11	56-63	Storm. Massive waves occur. Visibility is reduced to near zero.
12	64 or more	Hurricane. Air is filled with spray that blows horizontally. Huge waves occur.

ment (*anemometer*), but you can get a good idea of the wind speed by observing what it does to trees, dust, buildings, or the surface of a large body of water. Table 2-1 shows the Beaufort scale for winds observed on land, and Table 2-2 shows the Beaufort scale for winds observed at sea or on large lakes. The numbers range from 0 (calm) to 12 (hurricane force).

WIND FORCE

Perhaps you have wondered why a 40 kt wind does little or no damage, while an 80 kt wind, only twice as fast, can snap trees off at the trunk, shatter window panes, and send large objects flying through the air. Moving air produces measurable forces against objects that get in the way. This force increases much more rapidly than the speed in knots. This is why incredible things happen in severe hurricanes and in tornadoes. The power of the wind in a tornado can be thousands of times the power of a late-autumn gale that rattles the windows and strips the last of the leaves from the trees.

When a strong wind blows against a building, force is produced directly against the wall or walls facing most nearly into the wind. This is *positive force.* There are also other forces generated by winds blowing around an object. As the air flows over the roof, the pressure above the surface goes down, producing *negative force* as the air inside the building pushes upward. If the wind gets strong enough, part or all of the roof can be ripped off because of this force. Similar negative forces are produced on walls that face sideways to the wind; windows are sometimes blown out by this pressure. Some negative pressure also occurs on the wall or walls facing away from the oncoming wind.

PROBLEM 2-6

Suppose you look at an Internet site for tropical weather and learn that a hurricane has sustained winds of 120 kt. What is the equivalent wind speed in miles per hour? In kilometers per hour?

SOLUTION 2-6

From the above discussion, we know that a wind speed of 1 kt is equal to approximately 1.151 mi/h. Therefore, 120 kt = 138 mi/h. A speed of 1 kt is equal to approximately 1.852 km/h. Therefore, 120 kt = 222 km/h.

Clouds

Clouds normally form in parcels of air where the relative humidity is 100%, that is, when the air is saturated with water vapor. The temperature at which clouds form depends on how much water vapor is in the air. The more water vapor the air contains, the higher the temperature at which clouds can form.

CAUSES OF CLOUDS

The temperature of the air decreases steadily with increasing altitude, but the amount of water vapor does not necessarily decrease with altitude. Thus, the relative humidity typically rises with altitude. If it rises to 100% at a certain level, clouds can form at and above that level, because condensation forms on airborne dust particles. The development of clouds is accelerated by atmospheric updrafts, when moist air ascends to great heights. The higher the relative humidity, in general, the lower the altitude at which clouds begin to form. On a muggy or rainy day, clouds are usually observed at low altitudes, and in the extreme case, they form on the ground. Then they are called *fog*.

Meteorologists and aviators speak of the *cloud ceiling* (often simply called the *ceiling*). This is the altitude of the bottom of the lowest layer of clouds, measured with respect to either sea level or the earth's surface. Clouds form in a variety of shapes and patterns. Meteorologists have several different classifications for tropospheric clouds. Clouds are named according to their relative altitude and their general shape. Clouds associated with weather at the surface occur at altitudes from sea level to the top of the troposphere. *Nacreous clouds* and *noctilucent clouds* form above the troposphere, and are not directly associated with weather at the surface.

HIGH-ALTITUDE CLOUDS

The highest tropospheric clouds range from altitudes of about 7 km (20,000 ft) to 18 km (60,000 ft) above sea level. They are named with words that begin with the prefix *cirr-* from the Latin word cirrus, meaning "curly." Wispy, web-like *cirrus* clouds are commonly seen. High, smooth clouds are called *cirrostratus.* They give the sky a milky appearance and frequently cause a ring around the sun or moon.

An old adage says that a ring around the sun or moon means rain or snow is on the way. This is often true. Low-pressure systems are commonly preceded by high, thin clouds such as cirrostratus. Some high clouds have a congealed or puffy look. These are known as *cirrocumulus.* They can cause partial or complete overcast.

MID-ALTITUDE CLOUDS

The mid-level clouds are between about 800 m (6000 ft) and 7 km (20,000 ft) above sea level and are given the prefix *alto-*, which is Latin for "high."

Mid-level clouds sometimes have a flattened appearance, in which case they are called *altostratus.* Altostratus clouds are lower and thicker than cirrostratus. Mid-level clouds with a puffy look are known as *altocumulus.*

LOW-ALTITUDE CLOUDS

The lowest clouds have bases at altitudes less than 800 m (6,000 ft). These clouds are sometimes given the prefix *strat-*. Low clouds sometimes extend all the way to the surface, in which case *fog* occurs.

Low, flat clouds are called *stratus* clouds. If rain falls from them, they are called *nimbostratus.* Low, rolling, gray *stratocumulus* are seen on an overcast but dry day typical of much of the United States in late autumn. The cottonlike puffs of fair-weather *cumulus* are easy to recognize.

CLOUDS AT MULTIPLE ALTITUDES

Several kinds of clouds can exist at the same time, at various altitudes. It is not uncommon, especially in storm systems, to observe three different and distinct cloud layers. One well-known type of cloud complex, the *cumulonimbus,* encompasses low, medium, and high levels, often extending from approximately 1 km (3300 ft) to the top of the troposphere (Fig. 2-9). These clouds sometimes produce severe thunderstorms.

Cumulonimbus clouds sometimes stand alone in an otherwise clear sky. In other storm systems, especially in front of strong cold fronts, multiple cumulonimbus clouds merge into *squall lines.* Cumulonimbus clouds congregate in a donut-shaped, revolving mass around the eye of a tropical storm or hurricane.

Imagine that it is a cloudy, dark day, and you board an airplane bound for a distant place. You know the airplane will be flying at a high altitude. As you settle in for the long ride, the pilot's voice comes over the speakers. She tells you that the cruising altitude will be 37,000 feet (approximately 11 km or 7 mi). The plane climbs into a nimbostratus cloud bank. The clouds are so thick that you can hardly see the wings. Then the aircraft bursts into the clear, only to reveal another, higher cloud layer. You recognize it as altocumulus. Even at this altitude, the sky is overcast. The plane climbs higher and enters the second deck of clouds. You emerge from the second cloud deck into sunshine, but you can see clouds in the distance at a still higher altitude. As you squint into the sun, you discern a cumulonimbus "thunderhead" about 60 km (40 mi) away, and cirrostratus at about the same distance. The altocumulus clouds, through which you

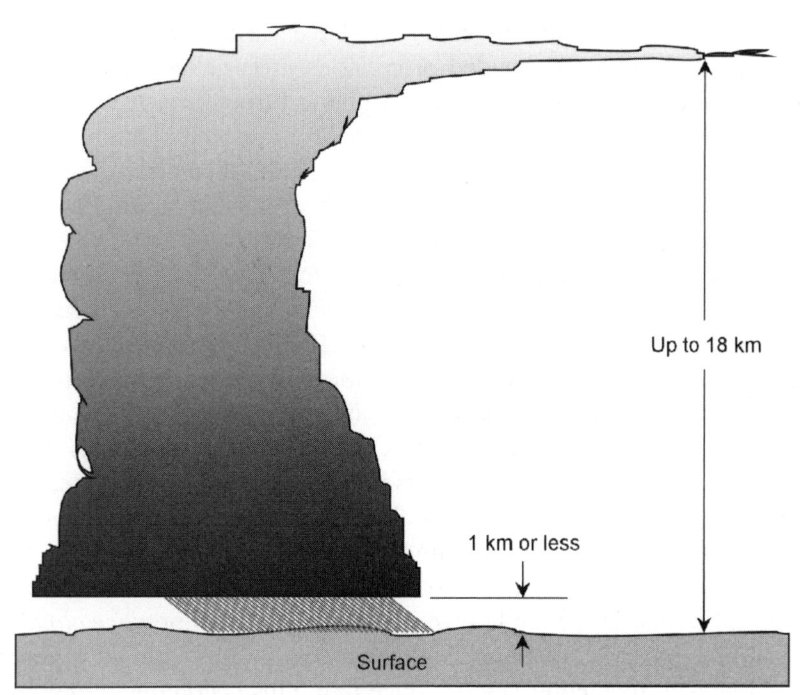

Fig. 2-9. Cumulonimbus clouds affect the troposphere at all altitudes, and often occur along with other cloud types.

have just passed, recede underneath the aircraft, which is still gaining altitude. This scenario is typical of midsummer over continents at temperate latitudes.

OTHER CLOUD TYPES

Clouds do not always fall neatly into common classifications. There might be some doubt, for example, whether a certain cloud is cumulus or altocumulus. Not all clouds of a given type look exactly the same. The wind speed at the cloud level, the angle from which the sun or moon is shining, the extent of cloud cover at higher or lower altitudes, and other factors affect the appearance of clouds.

Some clouds are unusual or interesting enough to have special names. For example, when cumulus clouds build up in the heat of the summer, towering higher and becoming wider at the base, they are called *cumulus congestus.* The "thunderheads" nearly merge together. When rain begins to fall from such clouds, they technically become cumulonimbus.

The base of a cumulonimbus cloud can have various features. If the cloud seems to have pouches or rounded protrusions, it is called *mammatus.* These clouds are associated with extreme atmospheric turbulence. A cyclonic vortex will occasionally form within a cumulonimbus cloud, the spinning air column extending below the base. If the vortex is strong enough, water vapor condenses and a *funnel cloud* appears. If the funnel cloud reaches the ground or water surface, it becomes a *tornado* or *waterspout.*

When high winds are present at relatively low altitudes, we sometimes see ragged, fast-moving, dark clouds called *fractocumulus* or "scud." These clouds are literally shreds of cumulus or cumulonimbus. They scurry along, driven by high winds, and are common in and near violent storms. Fractocumulus clouds are usually dark, because they hang in the shadows of higher clouds. But sometimes they contain dust and dirt, so they appear dark even when the sun shines directly on them. This is the result of strong updrafts in the past history of a storm.

Mountains can produce marked local effects in the atmosphere, at altitudes ranging from a few thousand meters up to several kilometers. As moist air passes over a mountain range, lens-shaped or undulating clouds often form on the leeward sides of peaks. These are known as *lenticular clouds* or *wave clouds.* They are a combination of altostratus and cirrostratus, although they sometimes become large and thick enough to produce rain, thus becoming cumulonimbus.

HUMANMADE CLOUDS

Not all clouds are products of nature. High-flying aircraft produce their own form of cloud: the *vapor trail* or *contrail.* A contrail can result from condensation of water vapor in jet-engine exhaust as the rarefied air cools it. A contrail can also be produced by the abrupt decrease in pressure that chills air as it passes over fast-moving airfoils. Rockets produce contrails as they ascend through the upper troposphere. If a vapor trail persists long enough, it is spread into cirrostratus or cirrocumulus clouds by high-altitude winds. In recent decades, an increasing number of aircraft have been flying in the upper troposphere, and some scientists think this has caused an increase in the average amount of cloud cover, especially over the midwestern United States.

Another example is the partially humanmade cloud, *smog,* that is found in and near major urban areas. The term is derived from a combination and contraction of the words "smoke" and "fog." There is little doubt that smog is bad for the environment, but there is disagreement about the extent of the environmental damage smog has caused on our planet.

Sometimes you will see what looks like a cloud, but really isn't. A forest fire can produce blue-white or gray smoke that looks like a cloud from a distance. Some such "clouds" can rise thousands of meters, blow downwind for distances in excess of 100 km, and show up on weather radar displays and satellite photographs.

In some places, brush fires can produce smoke that inhibits visibility and causes general darkening of the sky. In the late winter and early spring in South Florida, dry weather in conjunction with lightning storms or careless tourists can cause fires in the Everglades. If the wind blows from the west, as it often does during the winter and spring, the cities of the "gold coast"—Miami, Fort Lauderdale, and places between—are covered by dense smoke. It smells like burning leaves and looks like fog. At its worst, it can cause breathing problems for people with respiratory ailments.

CLOUDS FROM THE EARTH

Blowing dust or sand, as well as blown smoke, can take on the appearance of threatening clouds. Sand, smoke, or dust occasionally fill the air over vast regions. Massive forest fires in the western regions of the United States have caused smoke that rose to heights of several kilometers, drifted thousands of kilometers, and darkened the skies as far east as the Atlantic seaboard. Strong winds over a large area, during a period of dry weather, can pick up and carry the topsoil for hundreds of kilometers, blackening the sky over a vast region.

Some clouds of debris are produced by volcanic eruptions. After the eruptions of Mount St. Helens in 1980, for example, much of the United States was affected by the volcanic dust that rose to the top of the troposphere and was caught and carried by the jet stream. Volcanic dust has been blamed for cooling trends lasting several years. Some scientists believe that the legendary cold summer of 1816 and the cold winter of 1983–1984 were caused by a prolonged reduction in the amount of solar heat received by the earth, and that this heat reduction was the direct result of blockage and reflection of solar energy by airborne volcanic dust. In both cases, massive volcanic eruptions months earlier threw tremendous amounts of dust into the upper troposphere and stratosphere.

CLOUDS WARN OF STORMS

Certain clouds, or combinations of clouds, indicate the approach of foul or severe weather. Other clouds reassure us that the weather will stay fine for awhile. You can usually get a good idea of what to expect, simply by observing clouds.

Small, puffy cumulus clouds, with little or no clouds at higher altitudes, generally mean that the weather will remain fair for the next day or so. If the clouds begin to build up during the late morning or early afternoon, however, you should take your umbrella if you plan to be outside later. Gradually rising and thickening cumulus do not usually portend severe weather, although heavy rains sometimes occur. This kind of weather pattern is typical in the tropics and subtropics during much of the year, and in the temperate latitudes in the summer.

A distant, but advancing, storm system causes increasing instability in the upper troposphere. A clear, deep-blue sky, in which airplanes produce no contrails or contrails that die out quickly, indicates stability in the upper troposphere, and the continuance of fair weather for a day or two. The first noticeable sign of an approaching low-pressure system is the tendency for aircraft to produce long contrails that span the sky from horizon to horizon.

The upper troposphere gets increasingly susceptible to cloud formation as water vapor condenses more readily, and this tends to happen in advance of an approaching storm system. As the system moves nearer, high cirrus clouds appear. They get thicker until the sky is filled with high clouds. At temperate latitudes, the clouds thicken toward the western horizon. The movement of the clouds is generally from west to east. In the tropics, clouds thicken toward the east; the cloud movement is usually from some point in the eastern half of the compass toward some point in the western half. As the atmosphere becomes more unstable, cumulus congestus clouds begin to form. The wind shifts, backing several compass points. In the temperate latitudes, the prevailing westerlies give way to breezes that come from the direction of the equator. In the tropics, the easterly trade winds shift, become gusty, and blow from the direction of the pole. Eventually, rain or snow begins to fall. At temperate latitudes, the temperature can rise several degrees Celsius over a period of a few hours. Then the sky lightens, the rain or snow abates, and moderate weather prevails. If the storm system is intense, however, these pleasant conditions do not last long. In the tropics, conditions are difficult to predict without satellite information. The approaching system might be nothing more than a weak tropical wave, but it might be a hurricane!

For those who live in the latitudes where frontal cyclones prevail, the sequence of events just described indicates the approach and passage of a warm front. When a warm front has moved past an area in the temperate zone, a cold front often follows. The cold front causes large cumulonimbus clouds to form in the summer. In the winter, the wind veers sharply and increases in speed, sometimes to gale force. If snow has fallen or already exists on the surface, it blows in a blinding white sheet. At any particular surface location affected by the front, the temperature falls suddenly and dramatically as the front arrives and the cold air mass behind it takes over.

As a cyclone approaches in the tropics or subtropics, changes in temperature are not likely to be significant. The sky gets alternately gloomy and bright. If the storm is not intense, rain falls, accompanied by gusty winds. If the storm is intense, the National Hurricane Center informs people in or near its forecast track. Before the existence of geostationary satellites and reconnaissance aircraft, people in the path of a hurricane had little or no warning before the high winds and tides were upon them.

As a low-pressure system moves away, the weather improves. Behind a temperate low, brisk winds blow from the pole, and the weather becomes dry. Within a few days, the pattern is repeated as another system approaches and passes. Behind a tropical storm, conditions gradually improve as the winds and rains ease and the trade winds regain their benign aspect.

PROBLEM 2-7

Why does the sky get so dark during a heavy thunderstorm?

SOLUTION 2-7

The "fog" of water droplets (which scatter sunlight) in a cumulonimbus cloud extends continuously from an altitude of a few hundred meters up to several kilometers. As a result, more sunlight is absorbed than is the case with layered clouds, even if the minimum-to-maximum–altitude cloud span is the same. The angle of the sun with respect to the horizon also has an effect. The larger the angle, the less cloudiness the sunlight must penetrate in order to reach the surface. A summer storm that occurs at noon (when the sun is high) does not cause as much darkening as is the case if it occurs in the early morning or late afternoon (when the sun is low).

PROBLEM 2-8

In recent years, much attention has been given to the damaging effects that sunlight has on exposed skin. Do clouds protect against ultraviolet (UV) radiation from the sun? If they do, why do we often hear about people getting sunburned on overcast days?

SOLUTION 2-8

Heavy overcast reduces the intensity of UV reaching the surface, as compared with clear weather. However, this reduction is not always as great as people imagine, so they tend to neglect precautions such as applying sunblock lotion to the skin, wearing long-sleeved shirts, long pants, and wide-brimmed hats, and limiting sun-exposure time. This is why some of the worst cases of sunburn occur on cloudy days.

Quiz

This is an "open book" quiz. You may refer to the text in this chapter. A good score is 8 correct. Answers are in the back of the book.

1. Low, dull, "iron gray" clouds that completely cover the sky, and from which a light rain falls, are an example of
   (a) cirrus.
   (b) cumulus congestus.
   (c) nimbostratus.
   (d) fractocumulus.

2. Clouds often form when
   (a) the air is suddenly heated.
   (b) air is compressed.
   (c) air is suddenly cooled.
   (d) All of the above

3. Fill in the blank in the following sentence to make it true: "The mid-latitude jet streams carry high and low pressure systems generally ________ at temperate latitudes, primarily between 30°N and 60°N, and between 30°S and 60°S."
   (a) from the equator toward the pole
   (b) from lower altitudes to higher altitudes
   (c) against the direction of the prevailing winds
   (d) from west to east

4. The troposphere is
   (a) the highest layer of the atmosphere.
   (b) the only layer in the atmosphere where no weather occurs.
   (c) also known as the torrid zone.
   (d) None of the above

5. A cumulonimbus cloud can rise as high as the top of the
   (a) troposphere.
   (b) stratosphere.
   (c) mesosphere.
   (d) thermosphere.

6. In the northern hemisphere, the term *backing*, with respect to wind, refers to
   (a) a counterclockwise shift in compass direction; for example, from west to south.
   (b) a clockwise shift in compass direction; for example, from north to east.
   (c) a decrease in speed.
   (d) a wind that blows contrary to the direction of the prevailing winds.

7. Which of the following is an indicator that stormy weather can be expected at a location in the temperate latitudes?
   (a) A gradually falling temperature reading.
   (b) A rapidly falling barometer reading.
   (c) A sudden wind shift to the north.
   (d) Gradually decreasing relative humidity.

8. The expression "knots" is equivalent to
   (a) statute miles per hour.
   (b) kilometers per hour.
   (c) nautical miles per hour.
   (d) meters per second.

9. The altitude of the base of the lowest clouds in a given location is called the
   (a) ceiling.
   (b) thermal level.
   (c) visual level.
   (d) fog level.

10. At certain levels in the atmosphere, ultraviolet (UV) radiation from the sun
    (a) is changed into visible light.
    (b) causes ionization of atoms.
    (c) causes lenticular or wave clouds to form.
    (d) causes noctilucent clouds to form.

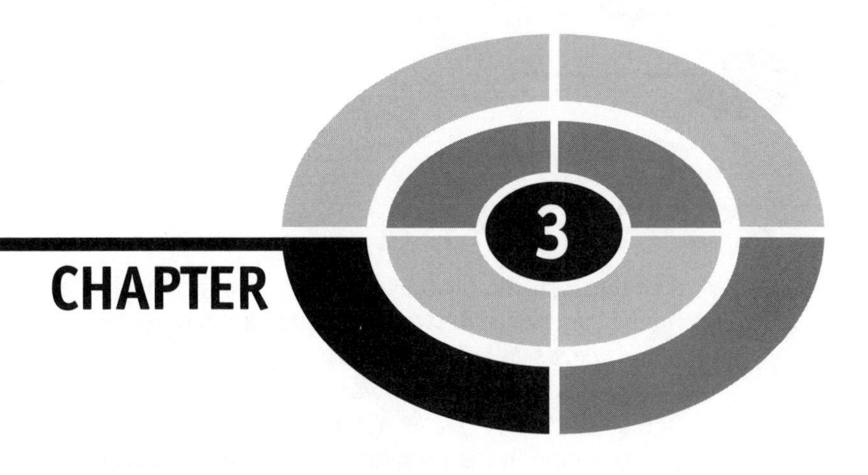

Observation and Forecasting

People have tried to forecast the weather for at least as long as there has been recorded history. All modern forecasting methods involve observations of current conditions, along with a combination of historical data, scientific method, and computer modeling.

Some Weather Lore

Before people had instruments to measure parameters such as temperature, humidity, and barometric pressure, there were ways in which the weather could be forecast—or at least, ways in which attempts at forecasting could be made. People noticed that certain observations or events were usually followed by fair weather, while other observations or events were usually followed by foul weather. Here are some weather legends, or bits of lore, that have evolved.

SILVER-SIDE OUT

"Before a storm, the leaves of the trees face backside to the breeze." Tough oaks in the American Midwest are called upon to withstand extreme temperatures and wind speeds. Whole hillsides are covered with them. As the wind blows from the south or southwest on a warm, humid day in late spring or summer, you'll see the back ("silver") sides facing outward.

A change in the humidity can cause leaves on the trees to alter the way they hang, because the stems absorb water in an uneven way. Storms are often preceded by a rise in the humidity, especially in the American Midwest and Northeast, where low-pressure systems pump warm, moist air into a region before the cold front, with its associated turbulence, passes through. This bit of lore has some science behind it!

THE WEATHER STICK

Long before Europeans came to North America, Native Americans in the region now known as New England discovered that the branches of certain trees bent in one direction in fair weather, and in the opposite direction in foul weather. These branches, measuring about 30 cm (1 ft) long, were de-barked, mounted in plain sight, and used as weather-forecasting instruments. They can still be obtained today under names such as *Maine Weather Stick.* You can find some of these for sale on the Internet by entering "weather stick" into any search engine.

There's a Maine Weather Stick mounted on the side of my house, and it can be watched from the dining room window. During fair weather it points up, and during foul weather it points down—usually. Upon close observation, I have found that warm temperatures and low humidity cause the stick to move upward, and cold temperatures and rising humidity cause it to move downward.

PAINFUL JOINTS

Have you ever heard anyone say, "My joints tell me that we're going to have a storm"? Maybe you have a body like this! Some people get headaches in association with weather changes; other people feel the same no matter what happens, and call all this talk about "rheumatism" and "migraines" a mark of a hypochondriac.

Changes in barometric pressure can cause some people to develop reactions in their bones and sinuses. The onset of sultry weather prior to a storm can cause some people to experience mood changes, and these in turn can affect the way they feel physically. Certain weather conditions can cause an increase or decrease in air pollution or pollen levels, in turn affecting the severity of allergic reactions.

As people age, their bodies get increasingly sensitive to variable weather. This has been well documented, and was described by Native Americans. So it's not some import from the pilgrims. However, each person reacts uniquely to the weather, and only experience can provide anyone with a "forecasting tool" that is "based in the bones." At best, it is a qualitative phenomenon, subject to large errors and variability.

BIRDS ROOST BEFORE A STORM

The notion that birds roost (sit around and don't fly much) when a storm is approaching has a sound basis, assuming birds can sense weather changes. During times of adverse atmospheric conditions, the informed small-plane pilot stays on the ground. Before a storm comes in from the sea, small craft remain in port. Do birds really have a built-in sense of the weather to come? Many people, especially rural folks, will tell you they do. If they do, then they apparently heed it. The next time a big thunderstorm comes near, watch the birds.

Similar behavioral changes have been observed among other animals. During the Klondike gold rush, sled dogs were known for digging their sleeping holes in the snow on the sides of trees that later faced away from the wind during a blizzard. Some ranchers say they can tell whether the next day will be fair or foul, based on the behavior of their cattle. In the Florida Keys, some natives say they can tell when a hurricane looms, because the roaches get restless.

CLOUDY BEFORE SEVEN

When I was a child, one of my teachers said, "Cloudy before seven, clear before eleven," and its converse, "Clear before seven, cloudy before eleven." According to these principles, clear nights should always give way to cloudy days, and gloomy dawns should invariably evolve into sunny afternoons. Perhaps these axioms apply in certain places at certain times of the year, but whoever coined them never spent a winter in the woods of Wisconsin or a summer in the high desert of Nevada.

STORING UP FOR WINTER

Here's another bit of country weather lore. Animals develop thicker fur, put on more fat, or store away more food prior to a long or severe winter, than they do before a short or mild winter. If you see squirrels storing acorns in greater abundance than usual around Labor Day, or if you see your "outside cats" getting fat and fuzzy for Halloween, you should suspect, according to this theory, that the approaching winter will be cold or snowy, or both, and will linger into May—but only if you live up North or in the mountains, of course. People in the Sun Belt (the southern U.S.) don't have to worry about this.

Anyone who has lived both the country life and the urban life, and who knows anything about animals, knows that their behavior correlates with the weather. Humans, in developing the intellect, have sacrificed instinct and intuition. But animals, like humans, can get things wrong. I've seen winters that started out severe become mild in December and then stay mild. I've also seen winters that withheld snow until January turn stormy and frigid, lasting well past the spring equinox. I wonder if anyone else has noticed this? Or can we dismiss it as a trait peculiar to *Homo sapiens*—imagination?

THE LAST WORD: SCIENCE!

Animals, as well as trees and joints, aren't perfect weather forecasters. If taken literally, folklore forecasting can lead to serious consequences! That's why people have developed instruments for quantitative measurement of weather parameters, and put the power of scientific method to work in formulating weather and climate forecasts.

Today, when you see a big thunderstorm coming straight towards your town on Doppler radar by checking out the Web page for your Zip code at the Weather Channel (www.weather.com), you don't need to watch the birds, look at the trees, or flex your elbows. You *know* that rain will soon fall.

PROBLEM 3-1

How can a device such as the Maine Weather Stick actually work? Or is it nothing more than a gimmick?

SOLUTION 3-1

The wood in the top of the stick behaves differently—under changing temperature and humidity conditions—than the wood in the bottom of the stick. Fig. 3-1 illustrates the principle. The stick's position during good or improving weather is shown at A; the stick's position during bad or deteriorating weather is shown at B. There are at least four possible scenarios that can cause the stick to work.

- The wood in the top of the stick (X) tends to contract when the temperature rises and/or the humidity drops, conditions typical of improving weather; it tends to expand when the temperature drops and/or the humidity rises, conditions typical of deteriorating weather. The wood in the bottom of the stick (Y) does not expand or contract as the humidity and temperature change.
- The wood in the bottom of the stick (Y) tends to expand when the temperature rises and/or the humidity drops, conditions typical of improving weather; it tends to contract when the temperature drops and/or the humidity rises, conditions typical of deteriorating weather. The wood in the top of the stick (X) does not expand or contract as the humidity and temperature change.
- Both of the above are true.
- Other factors, such as changes in the barometric pressure, affect the wood. These factors could operate in addition to, or instead of, temperature and humidity changes.

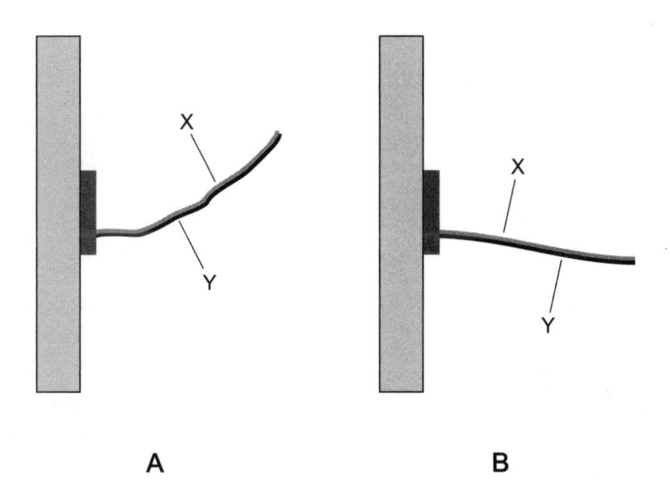

Fig. 3-1. Illustration for Problem 3-1. At A, the Maine Weather Stick during fair weather; at B, the same stick during foul weather.

Basic Observation Tools

These days, we have precision instruments that can measure parameters such as temperature and humidity. Some of these devices operate on principles similar to those of the Maine Weather Stick. Others are more sophisticated. Here are some examples of devices that can measure temperature, humidity, barometric pressure, and other variables. All of these should be included in a home weather station, if you're interested in observing the weather, recording daily conditions, or conjuring up your own forecasts.

THERMOMETER

Thermometers come in two types: the *bulb thermometer* and the *bi-metal-strip thermometer*. Either type can be used in a home weather station.

The bulb thermometer consists of an enclosed glass tube with a solution of colored liquid inside, and a bulb filled with the liquid at one end. As the temperature rises, the liquid expands, and the column of liquid in the tube gets longer. As the temperature falls, the liquid contracts, and the column of liquid in the tube gets shorter. A graduated linear scale indicates temperature. This type of thermometer has an advantage: It allows for two scales, one on either side of the column. Usually, when there are two scales, one is in degrees Fahrenheit (°F) and the other is in degrees Celsius (°C). Fig. 3-2A is a functional drawing of this type of thermometer.

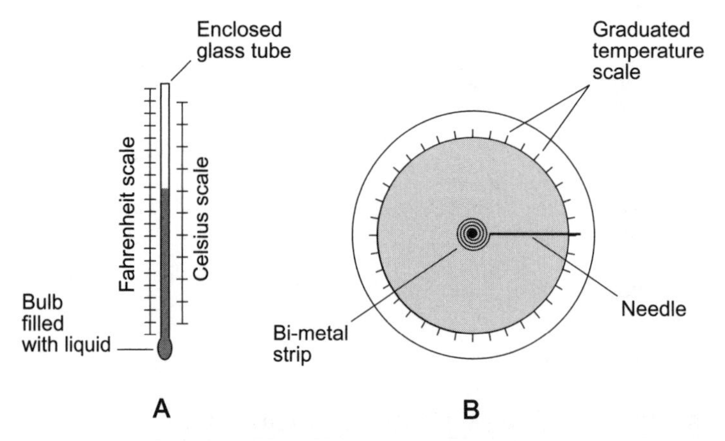

Fig. 3-2. At A, a bulb thermometer with dual temperature scales. At B, a bi-metal-strip thermometer.

The bi-metal-strip thermometer consists of two strips of metal with different coefficients of linear expansion, wound into a spiral. A needle is attached to the outer end of the spiral (Fig. 3-2B). As the temperature changes, one of the metal strips expands or contracts more than the other, and this changes the pitch of the spiral, in turn causing the needle to point in different directions. A graduated circular scale indicates temperature. The mechanism in this type of thermometer is also commonly used in *thermostats* that control older home heating and air-conditioning systems.

HYGROMETER

Relative humidity is measured by means of a *hygrometer.* The old-fashioned method of measuring relative humidity requires the use of two bulb thermometers, one with a "wet bulb" and the other with a "dry bulb." The rate of evaporation from the "wet bulb" depends on the amount of moisture in the air. Dry air causes more rapid evaporation than moist air, and hence a lower temperature reading. Relative humidity can be determined by taking the readings of both thermometers after they have stabilized, and then looking up the percentage of humidity on a table.

Simpler, but less accurate, hygrometers contain a human hair (or some other fine material that expands and contracts with changes in the air temperature and moisture content). This type of mechanism is found in hygrometers available in department stores. The hair is connected to a needle that moves back and forth, in a circle, or up and down along a calibrated scale. This type of hygrometer works according to a principle that is similar to that of the Maine Weather Stick—and the phenomenon responsible for "bad hair days"!

Still another way to measure relative humidity is by measuring the electrical conductivity of lithium chloride, a chemical salt. High levels of moisture in the air cause the lithium chloride to conduct better than it does when the air is dry. A battery, a milliammeter (a current-measuring meter), a resistor, and a container of lithium chloride salt can be connected in series to obtain an accurate hygrometer (Fig. 3-3)—once the current scale of the meter has been calibrated against the relative humidity under laboratory conditions.

BAROMETER

A *barometer* is a device for measuring atmospheric pressure. In most barometric pressure readings, a correction factor is introduced to standardize the actual air pressure to the sea-level equivalent pressure. This eliminates discrepancies

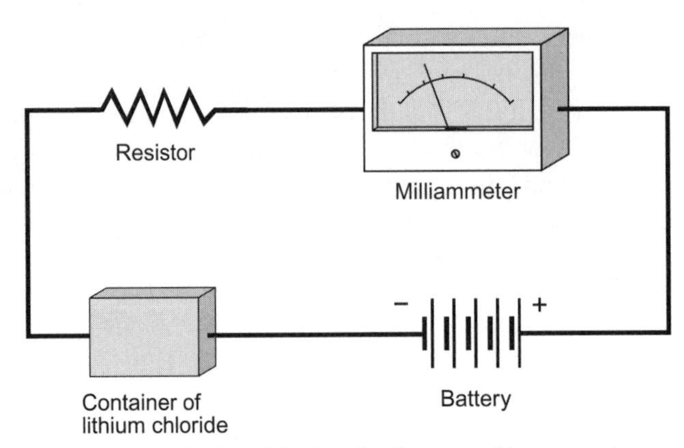

Fig. 3-3. Relative humidity can be determined by measuring the current through a sample of lithium chloride.

that would otherwise occur because of differences in altitude among weather observing stations.

In Chapter 2, the mercury barometer was described. This type of barometer, while accurate and reliable, is not widely available for the same reason you won't often find a mercury thermometer these days. Elemental mercury, as well as almost every compound derived from it, is toxic. Even the small amount of mercury found in an old-fashioned fever thermometer harms the environment when the device is discarded. A true mercury barometer requires more mercury than does a mercury thermometer, and is more destructive.

A common *aneroid barometer* has a sealed container called an *aneroid cell* from which the air has been partially removed. A needle is attached to the cell with a set of levers that amplify the needle movement as the cell expands and contracts under conditions of varying air pressure. The needle swings around a circular scale to provide a pressure reading. The scale is calibrated in pressure units called *inches of mercury* (inHg). A high-quality instrument is also calibrated in millibars (mb or mbar). The typical atmospheric pressure at sea level is a little over 1000 mbar.

In order to convert from inches of mercury to millibars, multiply by 33.864. If you want to convert from millibars to inches of mercury, multiply by 0.029530. Mathematically, if P_m is the sea-level barometric pressure in millibars and P_i is the sea-level barometric pressure in inches of mercury, then the following two equations hold:

$$P_m = 33.864\, P_i$$
$$P_i = 0.029530\, P_m$$

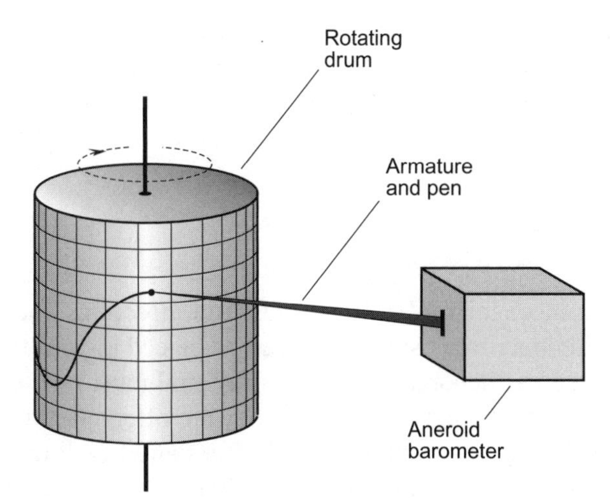

Fig. 3-4. An aneroid barometer can be attached to a pen recorder to get a graphical record of barometric pressure versus time.

The pointer in an aneroid barometer can be equipped with a marking tip that lies against a piece of quadrille paper attached to a rotating drum. The result is a *pen recorder* that can record barometric pressure over periods of hours or days. (Pen recorders can also be used with any other needle-type metering device.) The graphs produced by pen recorders can be interesting, particularly when they portray the passage of hurricanes or strong frontal cyclones. Fig. 3-4 is a functional diagram of a pen recorder connected to an aneroid barometer. Fig. 3-5

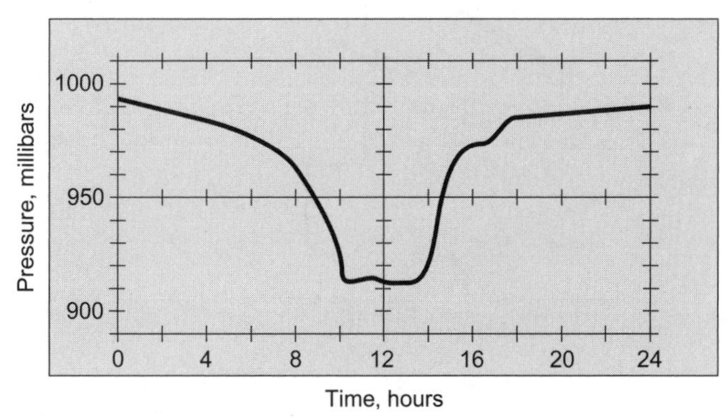

Fig. 3-5. Pen-recorder rendition of barometric pressure during the passage of a hurricane.

is a hypothetical example of the graph produced by a barometric pen recorder as an intense tropical hurricane scores a direct hit on the weather station where the recorder is located. The passage of the eye is represented by the flat interval of lowest pressure.

WIND VANE

A *wind vane* is a mechanism that indicates wind direction. The simplest wind vanes are found in department stores, and can be placed in your front yard. Most of these are intended as decorative devices only, and must be observed directly. They are not particularly accurate, because they are often placed where the wind direction is modified by obstructions such as trees and buildings.

A precision wind vane is connected to a *selsyn,* which is an electromechanical indicating device that shows the direction in which an object is pointing. It consists of a direction sensor and transmitting unit at the location of the movable device, a receiving unit and direction indicator located in a convenient place, and an electrical cable or wireless link that connects them. Fig. 3-6 is a functional diagram of a selsyn for use with a mechanical wind vane.

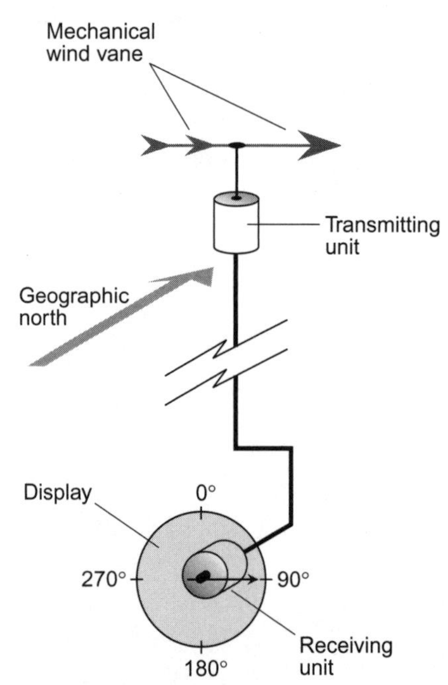

Fig. 3-6. A selsyn can be used with a mechanical wind vane to monitor the wind direction remotely.

In a true selsyn, the indicator rotates through the same number of angular degrees as the moving device. A selsyn for *azimuth* (compass) bearings, such as is used with a remote-reading weather vane, is calibrated in degrees clockwise from 0° (geographic north) through 90° (east), 180° (south), 270° (west), and all the way up to 360° (north, the same as 0°).

ANEMOMETER

An *anemometer* is a device that is used to measure wind speed. There are several types. One of the most common and well-known types has a set of three or four rotating cups. Another type has a propeller-like assembly that resembles a miniature wind turbine. Still another type employs a flat surface that hangs by a hinge, and that is deflected upward more or less depending on the pressure of the wind against it.

The rotating-cup anemometer (shown in Fig. 6-1, Chapter 6) operates independently of the wind direction. Other types are direction-sensitive, and must be attached to wind vanes in order to give accurate readings. A typical anemometer has a small electric generator connected by wires to an electrical meter that can be placed in a convenient location. The meter is calibrated in knots (kt) and/or miles per hour (mi/h). In some instruments, the meter is also calibrated in meters per second (m/s).

A typical anemometer can register wind speeds with reasonable accuracy up to about 65 kt, or minimal hurricane force. Beyond that speed, accuracy diminishes. Peak gusts are difficult to measure, especially with rotating-cup anemometers, because the mass of the assembly produces *angular momentum* that tends to keep the rate of rotation constant, resisting momentary changes in wind pressure. For this reason, particularly in severe storm systems such as hurricanes, the *sustained wind speed* is specified rather than the peak gust speed. The sustained wind speed is based on a 1-min average. Peak gusts in windstorms are typically 25 to 30 kt (30 to 35 mi/h) higher than the sustained wind.

Specialized anemometers are used in portable weather observatories that are deliberately placed in the paths of violent storms, particularly tornadoes. These devices are designed to withstand extreme winds that would destroy ordinary anemometers.

RAINFALL MEASUREMENT

A *rain gauge* measures the amount of liquid precipitation that has fallen between the present moment and the last time it was emptied. Normally, rain gauges are checked and emptied every 24 hours, except during storms when they can be checked more often.

A typical rain gauge has a funnel that empties into a test-tube-like container that is calibrated in centimeters and/or inches. The funnel increases the amount of rain that the device can "catch" (that is, it increases the *aperture*). When the liquid passes into the cylindrical portion of the device, the increments are expanded, making it possible to read rainfall amounts accurate to a small fraction of a centimeter or inch (±0.025 cm or ±0.01 in is typical).

The readings of a rain gauge can be influenced by various factors. High wind reduces the amount of rain that falls into the funnel, because the rain tends to blow across it, rather than down in. Special rain gauges are needed to accurately measure the amount of rain that falls in windstorms, particularly tropical cyclones. If a rain gauge is not emptied often enough, some of the water evaporates, producing an artificially low reading. During a significant storm, a rain gauge will overflow if it is not emptied often enough.

SNOWFALL MEASUREMENT

For measurement of snow accumulation, two figures are determined: the actual amount of snow as it has fallen on a flat surface (absent drifting or blowing), and the amount of liquid that remains when the snow is melted. The ideal surface for measuring snow depth during and after a snowstorm is an outdoor tabletop. This minimizes settling that can occur on warm ground when snow melts partially as it falls. Simply place a ruler or yardstick down into the snow, making sure it is oriented vertically, and making sure that the zero point on the ruler or yardstick is right at the end of the stick.

To measure the melted snow equivalent, a variety of techniques can be used. The simplest method is to collect a cylindrical sample of the snow from a flat surface, melt it down, and then measure the depth of the resulting pool of water. If this depth is too small to accurately measure, or if greater accuracy is desired, the contents of the cylindrical collecting container can be poured into a rain gauge, and the reading from the gauge multiplied by the ratio of the square of the diameter of the top of the rain gauge (that is, the funnel opening) to the square of the diameter of the cylindrical snow collector. If d_c is the diameter of a cylindrical snow collector, d_r is the diameter of the top of the rain gauge, and s_r is the reading of the rain gauge (in centimeters or inches) after the melted snow has been poured into it, then the melted snow equivalent, s_m (in centimeters or inches, respectively) is:

$$s_m = s_r d_r^2 / d_c^2$$

PROBLEM 3-2

Suppose you look at a weather map and see an isobar (a curve denoting points of equal barometric pressure) labeled 1012. This is the barometric pressure in millibars. What is the equivalent barometric pressure in inches of mercury?

SOLUTION 3-2

Use the formula for conversion from pressure in millibars (P_m) to pressure in inches of mercury, (P_i). We know that $P_m = 1012$. Therefore:

$$\begin{aligned} P_i &= 0.029530\, P_m \\ &= 0.029530 \times 1012 \\ &= 29.88 \text{ inHg} \end{aligned}$$

PROBLEM 3-3

Suppose there has been a snowstorm and you collect some of the snow in a coffee can by placing the can, open-side down, on a tabletop where there is snow. The can is 150 mm in diameter. You then slide the can off the top of the table, and use a piece of cardboard to keep snow from falling out. You take the can indoors and let the snow melt. Then you pour the water from the can into a rain gauge whose top is 200 mm in diameter. The rain gauge tells you that 0.45 in of "rain" has "fallen." What is the melted snow equivalent precipitation?

SOLUTION 3-3

First, determine the ratio of the square of the diameter of the top of the rain gauge (d_r^2) to the square of the diameter of the cylindrical container (d_c^2):

$$\begin{aligned} d_r^2/d_c^2 &= 200^2/150^2 \\ &= 40{,}000/22{,}500 \\ &= 1.77778 \end{aligned}$$

Then the amount of melted snow equivalent precipitation (s_m) is equal to 1.77778 times the reading of the rain gauge (s_r). Note that the multiplier 1.77778 is a *dimensionless constant,* meaning that it does not have units. We have been given s_r in inches. Therefore, the value we get for s_m will also be in inches:

$$\begin{aligned} s_m &= s_r d_r^2/d_c^2 \\ &= s_r \times 1.77778 \\ &= 0.45 \times 1.77778 \\ &= 0.80 \text{ in} \end{aligned}$$

If you hear this on the local weather forecast, they'll read it as "eighty hundredths of an inch."

Advanced Observation Tools

Professional meteorologists use sophisticated systems to aid in their observation of weather phenomena. The data and displays from these systems have become available to lay people in recent years through Internet sites such as the Weather Channel (www.weather.com).

RADAR

The term *radar* is an acronym derived from the words "radio detection and ranging." *Electromagnetic* (EM) *waves,* having frequencies in the *ultra-high* or *microwave* range, reflect from certain types of objects and particles, including aircraft, missiles, raindrops, ice pellets, and snowflakes. By ascertaining the direction(s) from which radio signals are returned, and by measuring the time it takes for an EM pulse to travel from the transmitter location to a target and back, it is possible to locate flying objects and to evaluate some weather phenomena.

A complete radar set consists of a radio-frequency (RF) *transmitter,* a *directional antenna,* an RF *receiver,* and a *display.* The transmitter produces EM wave pulses that are propagated in a narrow beam. The waves strike objects at various distances. The greater the distance to the target, the longer the delay before the reflected signal, or *echo,* is received. The transmitting antenna is rotated so that the sky in all azimuth directions can be observed. Most radar sets are aimed at the horizon, or a few degrees above it, so they can return echoes from the greatest possible distances.

In original radar designs, there was a circular display consisting of a *cathode-ray tube* (CRT). Nowadays, computers can take the signals from a radar receiver and render them on a high-resolution *liquid-crystal display* (LCD) or *plasma display.* The basic display configuration is shown in Fig. 3-7. The observing station is at the center of the display. *Azimuth bearings* are indicated in degrees clockwise from true north, and are marked around the perimeter of the screen. These show the compass direction of the echo relative to the station. The distance, or *range,* of the echo is indicated by the radial displacement of the echo. The radius on the display is directly proportional to the distance to the echo. Airborne

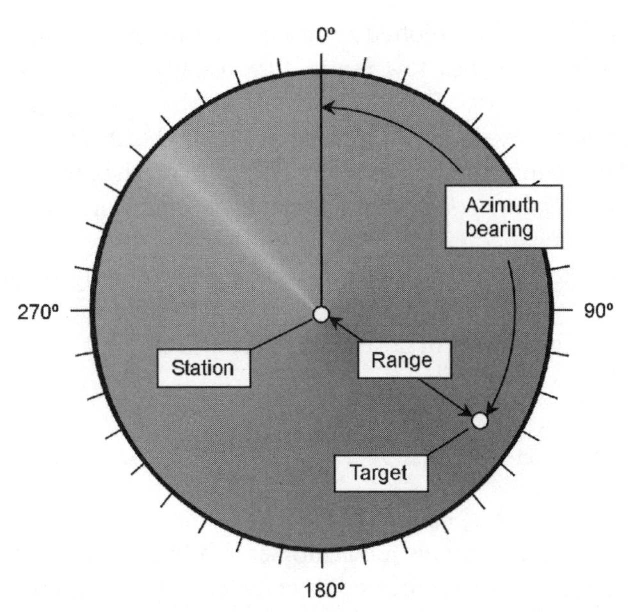

Fig. 3-7. A radar display. The light radial band shows the azimuth direction in which the RF beam is currently transmitted and received. (Not all radar displays show this band.)

long-range radar can detect echoes from several hundred kilometers away under ideal conditions. A modest system with an antenna at low height can receive echoes from up to about 80 km (50 mi) away.

Some radar sets can detect changes in the frequency of the returned pulses, thereby allowing measurement of wind speeds in hurricanes and tornadoes. This is called *Doppler radar.* This type of radar is also employed to measure the speeds of approaching or receding targets. Doppler radar has proven invaluable in detecting the presence of rotation within thunderstorms. This can provide advance warning of tornado development.

LIGHTNING STROKE DETECTION AND LOCATION

All lightning strokes produce bursts, or pulses, of EM energy. You can hear these as "static" on a portable radio tuned to the amplitude-modulation (AM) broadcast band. One method that can be used to locate a lightning stroke involves measuring the time it takes for the *EM pulse* from a given stroke to arrive at two or more receiving stations. This is a process familiar to navigators, and is known as *triangulation.*

Lightning strokes, when plotted on a regional map for a specific period of time (such as within the last few hours), are usually concentrated in and near areas of heavy precipitation. But this is not always the case. Some summer lightning in the United States desert Southwest occurs in the absence of precipitation. The rain in this region often evaporates before it reaches the surface, even in fairly strong thundershowers, because the air is dry and warm. Rain that does not reach the surface is called *virga.*

Lightning stroke locators are useful for people planning outdoor activities. They can also be useful to firefighting personnel in forested regions, because lightning is often the culprit in destructive forest fires.

INSTRUMENTS SENT ALOFT

A *radiosonde* is a portable, battery-powered weather station that is carried aloft by a helium balloon. It contains a thermometer, a hygrometer, a barometer, and a radio transmitter. The transmitter sends encoded data to stations on the ground. The position of the device in three-dimensional space at any given moment can be pinpointed using triangulation from multiple surface-based stations. In this way, the temperature, relative humidity, and barometric pressure are determined as functions of altitude, and also as functions of the point on the surface directly below the radiosonde (the exact geographical location). The wind direction and speed at various altitudes can be determined by observing and recording the direction and speed of the point on the surface directly below the device. The radiosonde can operate in the troposphere and lower stratosphere—as high as a balloon can take it.

A *dropsonde* is an instrument and radio transmitter similar to a radiosonde. The difference is that the dropsonde is released from a high-flying aircraft rather than sent up from the surface. It has a parachute instead of a balloon. A dropsonde is especially well suited to use in hurricanes and large thunderstorms. It can monitor conditions, and record wind speeds and the intensity of updrafts and downdrafts, at points within such storms that are too dangerous for aircraft.

Other vehicles for carrying instruments into the atmosphere include manned aircraft, helicopters, and rockets. Rockets can attain higher altitudes than other vehicles, so they can send back data at all levels of the atmosphere, not only the troposphere and lower stratosphere. The *hurricane hunter* aircraft is flown directly into the eyes of hurricanes, recording and observing conditions along the way. This aircraft is piloted by people who not only obtain quantitative data, but who directly observe (and vividly recall) qualitative conditions. In this respect, the difference between a radiosonde or dropsonde and a hurricane hunter is like the difference between the Mars lander and a manned mission to the planet.

WEATHER SATELLITES

The first *weather satellite* that provided pictures of the earth and cloud formations in the troposphere was called *TIROS.* It was launched in April, 1960. The TIROS satellite was approximately 0.6 m (2 ft) high and 1.2 m (4 ft) wide, and shaped like a can of tuna. Since this satellite was deployed, many others have been put into various types of orbits so that meteorologists can keep constant watch on the weather everywhere in the world.

The first weather satellites were in low orbits, at altitudes of only 200 km (125 mi) or so, and took approximately 90 minutes to complete a single revolution around the earth. Today, most weather satellites orbit at much higher altitudes, and they take longer to orbit the earth. When a satellite is put into orbit at a distance of approximately 36,000 km (22,500 mi) above the surface, the orbital period is 24 hours. This is called a *geosynchronous orbit.* Satellites in geosynchronous orbits above the equator are known as *geostationary satellites,* because they are always above the same point on the surface. In order to be geostationary, a satellite must orbit directly above the equator, and must remain at exactly the correct altitude all the time. It must also be revolving around the earth in the same direction that the earth rotates (counterclockwise as viewed from high above the north pole).

The first TIROS satellites orbited the earth at an angle of 48° relative to the equator. To observe a certain part of the world, it was necessary to wait until the satellite came within a few hundred kilometers of that location. Places at latitudes north of 55°N, or south of 55°S, couldn't be observed. Later, as rocket technology improved, satellites were placed in orbits tilted at angles of nearly 90° relative to the equator. Such an orbit is called a *polar orbit.* As the earth rotates beneath a polar-orbiting satellite, different parts of the world come into view. Over the course of half a day, a composite picture of the whole planet can be obtained. Such a combination of small photographs, put together to show a large area, is called a *mosaic.*

Today's fleet of weather satellites can view nearly an entire hemisphere of the earth at once, or zoom in on a particular weather system or surface feature. There are cameras sensitive to various wavelengths, including visible light, infrared (IR), and water-vapor (WV) emission spectra. The cameras can be adjusted to obtain more or less magnification. As the magnification increases and all other factors are held constant, the *image resolution* (detail) increases.

REMOTE MONITORS

Geostationary weather satellites are equipped with radio receivers and transmitters. Weather information is collected from remote monitoring stations on the surface. These include rain gauges, tide gauges, ships, manned weather stations, and automatically controlled weather stations.

A single satellite can be used to compile all the data from thousands of different surface-based stations, both manned and unmanned, at frequent intervals. The satellite can get the information from a particular station by sending a command to the station. Some stations send information to the satellite whenever unusual or severe weather conditions or other natural events occur.

LINEAR INTERPOLATION

Weather data is collected at defined, usually fixed, locations on the surface. The number of locations may be large, but it is not infinite. Therefore, conditions at points not located at observing stations must be estimated. A common way to estimate intermediate data is by *linear interpolation.*

Suppose it is a warm spring day in Minnesota, and the temperature at the weather station in Minneapolis is 21°C. In Rochester, approximately 100 km to the southeast, the temperature is 19°C. The temperature in Cannon Falls, located 50 km southeast of Minneapolis and 50 km northwest of Rochester, can be estimated by interpolation as the average of the temperatures at Minneapolis and Rochester, or 20°C. This is because Cannon Falls lies near the midpoint of a straight line connecting the two weather stations from which data has been obtained.

Interpolation does not always work. In regions where the terrain is irregular, elevation affects the temperature. In some locations, small-scale weather phenomena can be dramatic. Interpolation is not a reliable way to estimate data in these kinds of situations. In the Black Hills of South Dakota, for example, particularly in the winter, a few hundred meters difference in elevation can produce amazing differences in temperature. This is especially true when an *inversion* occurs, and warm air overlies cold air. Two stations, at the same elevation and separated by 10 km, might report temperatures of 5°C, while another station, midway between them but 200 m lower in elevation, reports −10°C.

Suppose, on that warm Minnesota day, an intense cold front moves through the state from the northwest towards the southeast. As the front passes Minneapolis, the temperature drops to 12°C. In Rochester, the afternoon sun has warmed things up to 21°C; the front has not yet arrived there. Interpolation is not a good way to estimate the temperature in Cannon Falls in this scenario. If the front has not yet arrived in Cannon Falls, the temperature there might be 20°C. An hour later, it might be 13°C, even if the temperatures in Minneapolis and Rochester have not changed during that hour.

The use of linear interpolation is not confined to temperatures. It can be used for any variable, including relative humidity, wind speed, wind direction, barometric pressure, and percentage of cloud cover. When using linear interpolation, it is important to keep in mind the limitations imposed by localized weather anomalies, irregular terrain, proximity to water, and other variables. Also, it's worth noting that linear interpolation is useful only between points that are fairly close together (a few tens of kilometers or less). Thus, you can't interpolate the conditions in St. Louis, Missouri, based on known conditions in Miami, Florida and Seattle, Washington.

PROBLEM 3-4

Imagine a situation in which linear interpolation is reliable, such as an open prairie without weather fronts in the vicinity. Suppose the temperature in Happyton is 20.0°C and the temperature in Blissburg, 70 km away, is 22.0°C. Now consider Joyville, which does not lie on a straight line connecting Happyton and Blissburg, but is 70 km away from Happyton and 30 km away from Blissburg. Using linear interpolation, give an estimate of the temperature in Joyville.

SOLUTION 3-4

Refer to Fig. 3-8. Suppose you take a straight-line trip from Happyton to Joyville. This is 70 km. Then imagine traveling in a straight line

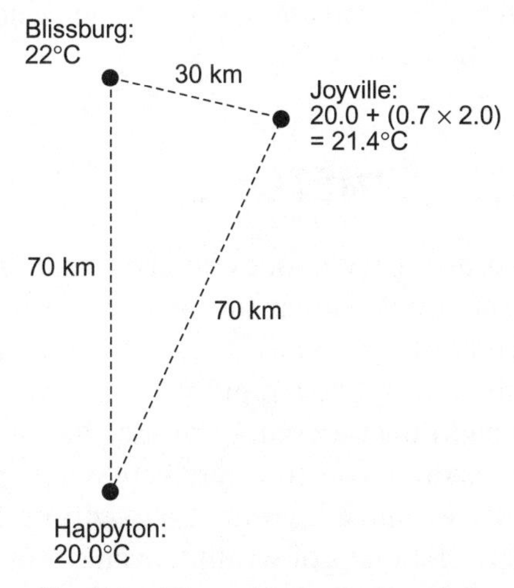

Fig. 3-8. Illustration for Problem 3-4.

from Joyville to Blissburg. This is 30 km. The total distance traveled is 70 km + 30 km, or 100 km. The temperature difference at the extreme points, representing the beginning and end of your complete trip, is 2.0°C. In terms of your route, Joyville is 70/100, or 0.7, of the way from Happyton to Blissburg. Thus, the temperature in Joyville can be estimated as $20.0 + (0.7 \times 2.0) = 21.4$°C.

WHAT DOES "LINEAR" MEAN?

In the foregoing scenario, the estimation process for the temperature at Joyville represents linear interpolation in the mathematical sense. But it is an imperfect application of the process, because the three towns don't lie on a geographical straight line. Linear interpolation works best when the location for which conditions are to be estimated lies on a geographical straight line between the two observation stations. If the evaluated location is not on this straight line, linear interpolation is less reliable. The greater the deviation from a straight line, the greater the chance for error in the result obtained by linear interpolation.

Forecasting Methods

The science of weather prediction has improved over the past several decades, largely as a result of the use of increasingly powerful computers and sophisticated observing equipment. Nevertheless, some basic principles, along with historical data, will always be useful.

HISTORY REPEATS—SOMETIMES

One of the simplest forecasting techniques involves examination of past weather events that followed conditions similar to those observed. For example, if you live in the American Midwest, you know that a warm, cloudy spring day with a southerly breeze, high humidity, and a rapidly falling barometer reading is more likely to produce a tornado than a cool, clear, dry day in early autumn with a high, steady barometer reading. If records are kept over a period of decades and then the data is carefully examined, correlations will be found. Certain conditions are followed by specific types of weather events—usually. But not always!

Forecasting based on history can be surprisingly accurate. It can also be astonishingly inaccurate! Have you ever seen a table that provides weather forecasts based on easily observed factors such as the barometric pressure and the wind direction? Table 3-1 is an example that can be used by amateur meteorologists to obtain forecasts in the American Midwest and Northeast. But this table does not work all the time. It won't work as well in Miami, San Francisco, or Honolulu as it will in Boston. It certainly won't be of much use to someone living in Sydney, Australia or in Santiago, Chile.

Table 3-1. Some common weather signs and forecasts for locations in the American Midwest and Northeast.

Barometer reading (inHg)	Barometer movement	Wind direction	Forecast
30.00 to 30.20	Steady	SW to NW	Fair for the next day or two.
30.00 to 30.20	Rising	SW to NW	Fair for a day or two; then rain or snow possible.
30.20 or above	Falling slowly	SW to NW	Fair for a day or two, with warming trend.
30.00 to 30.20	Falling	S to E	Becoming cloudy; rain or snow likely in a day or two.
30.00 to 30.20	Falling	S to NE	Rain or snow likely within 24 hours.
30.00 to 30.20	Falling	E to N	Rain or snow likely within 24 hours; rising wind.
30.00 or below	Falling slowly	SE to N	Rain or snow for a day or two.
30.00 or below	Falling rapidly	SE to NE	Rain or snow likely within hours; high wind possible.
29.80 or below	Falling rapidly	S to E	Heavy rain or snow likely within hours; high wind possible.
29.80 or below	Falling rapidly	E to N	Storm likely within hours; blizzard possible in winter.
29.80 or below	Rising rapidly	Veering to W or NW	Cooler, with rain or snow ending.

"Historical forecasting"—predicting the weather by assuming that what happened in the past will happen again this time—is one of the methods used to predict the tracks of hurricanes. Suppose you live in Charleston, South Carolina, and it is mid-September. You know that hurricanes can be a factor at this location during this time of the year. Suppose a category 3 hurricane (let's call it Ellen) is churning along just north of the island of Hispaniola. You go to one of the weather Web sites on the Internet and find a hurricane tracking map that shows the paths of all category 3 September hurricanes in the past several decades that have tracked within 320 km (200 mi) of Ellen at this point. You see that the tracks are concentrated in a generally west–northwesterly direction (approximately azimuth 290°) for some distance, and then they begin to diverge. By looking at these tracks, you can get some idea of the risk Ellen poses to you.

PERSISTENCE FORECASTING

One of the simplest methods of forecasting the weather is to assume that it will not change for a little while. This technique, called *persistence forecasting,* is often reliable over periods of minutes or hours. If a blizzard is blowing where you are right now, you can be pretty sure it will still be going 10 minutes from now, or 30 minutes from now. If the current barometric pressure is 29.65 inHg, you can be reasonably certain that it will be close to this value for the next few hours, unless a hurricane is bearing down on you.

In most locations, persistence forecasting doesn't work well for periods longer than a couple of hours, because the temperature and cloud cover often vary greatly between morning and afternoon, afternoon and evening, and of course, between day and night. There are some places, too, where it is risky even to attempt a forecast for the next half hour. Have you heard people say, "If you don't like the weather, wait a few minutes"? That can be literally true in locations such as the northern Black Hills in the summer, where the sun can be shining at one moment and marble-sized hail falling five minutes later. It is also true just before and during the passage of a "Texas norther," a rainband in a hurricane, or a fast-moving thunderstorm.

TREND FORECASTING

A more reliable and versatile method of short-term forecasting involves the assumption that a current tendency or trend will continue, or that an observed cycle will repeat. This is known as *trend forecasting.* It is a form of *extrapolation*—extending a defined rate of change into the future.

Suppose that it is 11:00 A.M. in your town, and you have been observing the temperature for the past several hours. At 8:00 A.M. it was 20°C. Then it rose to 22°C at 9:00 A.M., 24°C at 10:00 A.M., and 26°C at 11:00 A.M. A trend forecast would suggest that at 12:00 noon, the temperature will be 28°C.

There are limitations to this method. You can't keep extrapolating and conclude that within a couple of days, the air will be hot enough to boil water! In fact, if a cold front is approaching or you live in the tropics where rain showers are an almost daily occurrence during the afternoon hours, the trend will soon be reversed. By 5:00 P.M. it might be only 20°C again, after the front has passed or the shower has cooled things down.

Trend forecasting has been applied to long-term climate prediction. The controversy over *global warming*—Is it really happening, or not?—has been largely resolved by observing average temperature trends since about the year 1900. Based on this data, you'll hear scientists warn that if current trends continue, the entire planet will be several degrees Celsius warmer, on the average, a century from now, as compared with conditions today. They will also report trends in other factors such as the frequency and intensity of hurricanes, the ocean temperature, the distribution of rainfall, and changes in sea level.

STEERING

Have you heard that storm systems are "blown along" by upper-atmospheric currents, especially the jet streams? To some extent this is true, but it is an oversimplification. The jet streams, which you learned about in the last chapter, blow west–to–east in the upper troposphere along narrow, riverlike channels of air. There are two or three jet streams in the northern hemisphere, and counterparts in the southern hemisphere. The most significant of these are the mid-latitude jet streams. One of these meanders through the temperate zones between latitudes of approximately 30°N and 60°N. The other behaves in a similar manner between approximately 30°S and 60°S.

Low-pressure systems in the temperate zones tend to track along the jet streams, centered slightly poleward of them. (Hurricanes in the tropics follow atmospheric currents, too, but do not interact with the jet streams unless or until they enter the temperate zones.) Blizzard tracks can be predicted by watching the behavior of the jet streams. Individual fronts, and thunderstorm cells or snow squalls within these large cyclonic systems, however, tend to form, dissipate, and redevelop in a different way. Rather than following the jet stream, these smaller complexes are guided by the circulation around the main cyclonic system, or "low."

On the Eastern Seaboard and Gulf Coasts of the United States, residents hear the term *steering currents* in conjunction with hurricanes. These are not jet streams, but are less well-defined upper-tropospheric currents that tend to blow east–to–west in the tropics and subtropics. The National Hurricane Center issues updates for storms during the hurricane season, which runs from June 1 through November 30 in the North Atlantic, the eastern North Pacific, the Caribbean, and the Gulf of Mexico. If a hurricane threatens North America, the point of landfall is estimated by extrapolation on the basis of how the steering currents will cause the storm to behave over the next few hours or days. These steering currents are sometimes strong, causing hurricanes to follow well-defined tracks at speeds of 10 to 20 kt. The tracks of such storms are easier to predict than the tracks of hurricanes that encounter weak steering currents. When the steering current is not well defined or blows at 5 kt or less, it can be difficult to predict where a hurricane will track. Such storms challenge meteorologists, and produce gray hairs in veteran oceanfront residents who know what hurricanes can do.

SYNOPTIC FORECASTING

The term *synoptic* means "generalized," or "taking a view of the whole." *Synoptic weather forecasting* is done by assembling weather maps of large regions from observed and reported data at numerous stations (station models). The defining tool of synoptic forecasting is the weather map.

Synoptic forecasting evolved before computers were available to analyze weather data in high detail. A meteorologist might look at a sequence of weather maps showing conditions in the United States at intervals of a few hours, and deduce from it the conditions likely to exist for various locations a few hours, a day, or two days into the future. Fig. 3-9 is a simplified example of such a map, showing a high-pressure system (H) centered in the western United States, and a low-pressure system (L) centered in the southeast. Two hypothetical towns, Plainsboro in the Midwest and Surfsburg on the East Coast, are shown. The arcs with arrowheads represent the general wind circulation around the weather systems. The heavy, dashed line shows the position of the jet stream.

At the temperate latitudes, major weather systems tend to move generally from west–to–east at 20 to 30 kt. Based on this, even a meteorologist can come to the reasonable conclusion that the weather should be improving in Plainsboro and deteriorating in Surfsburg over the next 24 hours. High-pressure systems are usually associated with fair and warm weather, while low-pressure systems are associated with foul weather.

More detailed information can be obtained from Fig. 3-9 if we know the time of year. If it is January, Plainsboro might expect to see cool and clear weather for the next day or two, while Surfsburg can expect high winds and rain, fol-

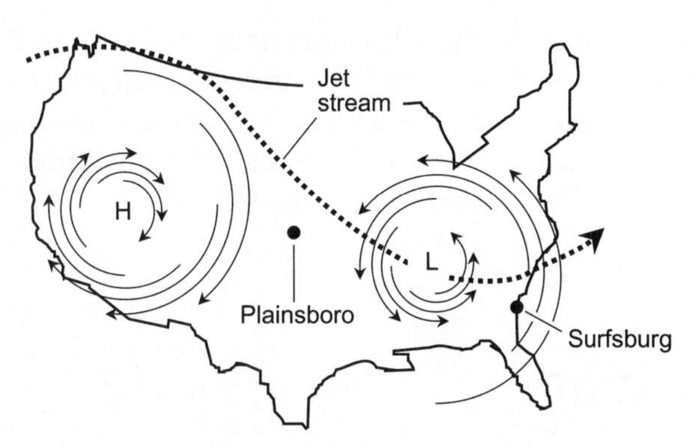

Fig. 3-9. A large-scale weather map of the United States, showing the jet stream, a high-pressure system (H) dominating the west, and a low-pressure system (L) dominating the east. Two hypothetical towns, Plainsboro and Surfsburg, are also shown.

lowed by cold weather and perhaps some snow. We would have to know the temperatures at numerous stations throughout the U.S. to get a better idea about this. If it is April, Plainsboro might enjoy a couple of warm, sunny days, while Surfsburg would brace for the possibility of severe thunderstorms or tornadoes. Again, the intensity of the weather systems, and the temperatures associated with them, would have to be determined by looking at the station models for numerous locations across the country.

Do you get the idea that synoptic forecasting is as much an art as a science? If so, you're right. Meteorologists in the early part of the 20th century relied on their experience and intuition, as well as on mathematics and physics. But that began to change in the 1960s and 1970s as computers became more available and more powerful.

PROBLEM 3-5

What can the town of Plainsboro and Surfsburg expect three days after the existence of conditions shown in Fig. 3-9? After a week? After two weeks?

SOLUTION 3-5

After three days, it's possible that the high-pressure system out west will have made its way to Surfsburg, giving them improved weather. But we can't be sure about this, based on the map alone. The storm system currently bearing down on Surfsburg will likely have moved out to sea. We have little idea of what to expect in Plainsboro after three days, because we must know what is coming in from the Pacific, and we must also know whether the high-pressure system out West will move along or remain fixed. The data in Fig. 3-9 is not enough to give us a good idea, much less a precise forecast, for conditions in either town after a week or two weeks have passed.

Numerical Forecasting

Synoptic forecasting can be compared to analog communications or imagery, in which variables can change continuously or "smoothly." Numerical forecasting is more like digital communications or imagery, in which the variables are broken up into small parcels or packets, and the behavior of the whole system is deduced by brute force of calculation, based on the laws of physics.

PARTICLES AND PACKETS

The atmosphere and the oceans interact to produce weather on a moment-to-moment basis. The motion of every single atom or molecule plays some role. The weather phenomena we see, and the conditions we experience, are the sum total of the actions of all these particles. There are millions upon millions of material particles moving around, and their interaction is complicated! But their number is finite, and each one of them obeys the same physical laws, which are clearly definable. It is impractical to observe the behavior of every atom and molecule in the planet's entire ecosystem, but the system can be broken down into parcels, and each piece considered an elementary particle.

Imagine that the atmosphere is defined according to a coordinate system that assigns cubic kilometers of air above square kilometers of planetary surface. Imagine that each square kilometer of surface has a stack of parcels measuring 1 km × 1 km × 1 km (for a volume of 1 km^3) of air above it, extending up to about 16 km (10 mi), the top of the troposphere. This stacking scheme doesn't work perfectly, because the earth is not flat, but 16 km is tiny compared with the radius of the earth, so it's almost perfect. Suppose that the water in the oceans is parceled in a similar way, considering cubic kilometers beneath the water surface, down to the seafloor, and that the same is done for large freshwater lakes. Now imagine that the characteristics of each parcel are measured by instruments located at the center of every cube. These conditions include:

- Temperature
- Relative humidity
- Wind direction (in three dimensions)
- Wind speed
- Barometric pressure

All of this data is fed into a computer, which is programmed with one objective: determining what each of these parameters will be, for each cubic kilometer of air or water on the entire planet, five seconds (5 s) into the future. It's not difficult to imagine that a powerful computer, programmed according to all the laws of physics, ought to be able to figure out a problem like that. This is, simplified terms, what numerical weather prediction is all about.

SMALLER PACKETS, LONGER TIME

Of course, a forecast for the next 5 s is not going to be of much use to anybody. But if we can forecast the weather 5 s in advance, we'll get a new set of parameters, and we can apply the same computer program to the same parcels of air again, getting another 5 s forecast. This can be repeated, or *iterated,* as many times as we like, ultimately getting forecasts for the next minute, hour, day, week, month, or year!

Obviously, as this numerical process is extended by repetition, the range of error increases. Even if we can predict, with 99.99% certainty, the parameters for every cubic kilometer of air and water for the next 5 s, we can't expect to get the same accuracy after 10 iterations (50 s), 100 iterations (500 s or 8.3 min), 1000 iterations (5000 s or 83 min), or 1,000,000 iterations (5,000,000 s or 58 days). The error, however small, that occurs with each iteration is multiplied by itself again and again. Over time, the resulting uncertainty can become so great as to render the forecast meaningless. If you multiply 99.99% by itself 1000 times, you get 90.48%. If you multiply 99.99% by itself 10,000 times, you get 36.79%. If you multiply 99.99% by itself 100,000 times, you get a number pretty close to 0%.

Yet, with each passing year, computers become more and more powerful, and the programs they use become better and better. Observing apparatus becomes more accurate, and can be placed in more and more locations. Imagine obtaining data, by a combination of observation and interpolation, for each parcel of air and water measuring just one cubic meter (1 m^3) in size! Suppose the accuracy is refined a thousandfold for every observation and every calculation! This might result in a 10 s prediction certainty of 99.9999999%. When this number is multiplied by itself 1000 times, you get 99.9999%. If you multiply 99.9999999%

by itself 10,000 times, you get 99.999%. If you multiply 99.9999999% by itself 100,000 times, you get 99.99%. Even if you multiply 99.9999999% by itself 100,000,000 times, you get 90.48%—a result as good as the 1000th iteration in the previous example.

IN PRACTICE

The preceding examples are hypothetical, and are provided only to give you an idea of how numerical forecasting can be done. Numerical forecasting has proven to be an excellent method of weather prediction for the short term (a few hours) and the medium term (up to a few days). Computer models have also been used to forecast weather trends for approaching months and seasons. Historical data can be compiled and analyzed, compared with current data, and extrapolated into the future.

Quantitative analysis of observed data, along with interpolation, extrapolation, and computer modeling, have provided scientists with fascinating insight into the future of the earth's climate and weather patterns. A good example is that, from a scientific perspective, there is little doubt that human activity is heating up the earth. A few scientists still question the theory of global warming, but not many.

The greatest challenge meteorologists and climatologists have ever faced can be posed as a question: How does the way we treat the earth today determine the way the earth will treat our grandchildren? And to this, we might add a second question: Even if we can generate a computer model that answers the first question with virtual certainty, will we be able to muster the collective will to take the appropriate action?

Quiz

This is an "open book" quiz. You may refer to the text in this chapter. A good score is 8 correct. Answers are in the back of the book.

1. A geostationary satellite always orbits
   (a) directly above the earth's equator.
   (b) at an angle of about 48° with respect to the earth's equator.
   (c) at an angle of about 90° with respect to the earth's equator.
   (d) above the meridian representing 0° longitude.

2. As the time frame for a numerical weather forecast increases, all other factors being constant, the extent of the error
   (a) does not change.
   (b) decreases.
   (c) increases.
   (d) becomes zero.

3. Numerical forecasting has been improving in recent years because
   (a) meteorologists have developed better intuition.
   (b) computers have become more powerful.
   (c) weather maps have become easier to read.
   (d) the weather itself has become less variable.

4. A method of weather forecasting in which it is assumed that current conditions will not change for a few minutes or hours is called
   (a) persistence forecasting.
   (b) linear interpolation.
   (c) numerical extrapolation.
   (d) iteration.

5. Suppose you are looking at a data table that tells you it was +5°F at 2:00 P.M. at the weather station in Wonderington, and +9°F at 6:00 P.M. at the same location. From this, you estimate that it was +7°F at 4:00 P.M. at the weather station in Wonderington. This is an example of
   (a) trend forecasting.
   (b) extrapolation.
   (c) synoptic weather prediction.
   (d) None of the above

6. Some hygrometers operate on a principle similar to that of
   (a) a mercury barometer.
   (b) an aneroid barometer.
   (c) an anemometer.
   (d) the Maine Weather Stick.

7. The electrical conductivity of certain chemical salts is sometimes used to measure
   (a) temperature.
   (b) wind direction.
   (c) relative humidity.
   (d) solar illumination.

8. The term "steering currents" is often heard in conjunction with
   (a) the circulation around a high-pressure system.
   (b) the air turbulence within a thunderstorm.
   (c) the difference in wind speed in various parts of a frontal cyclone.
   (d) hurricane track prediction.

9. The position of the jet stream can often be used to predict
   (a) where large low-pressure systems will go.
   (b) where small thunderstorms in larger cyclones will go.
   (c) the temperature difference between the upper and lower atmosphere.
   (d) the wind speed at any given location at any given time.

10. A polar-orbiting satellite always orbits
   (a) directly above the earth's equator.
   (b) at an angle of about 48° with respect to the earth's equator.
   (c) at an angle of about 90° with respect to the earth's equator.
   (d) above the meridian representing 0° longitude.

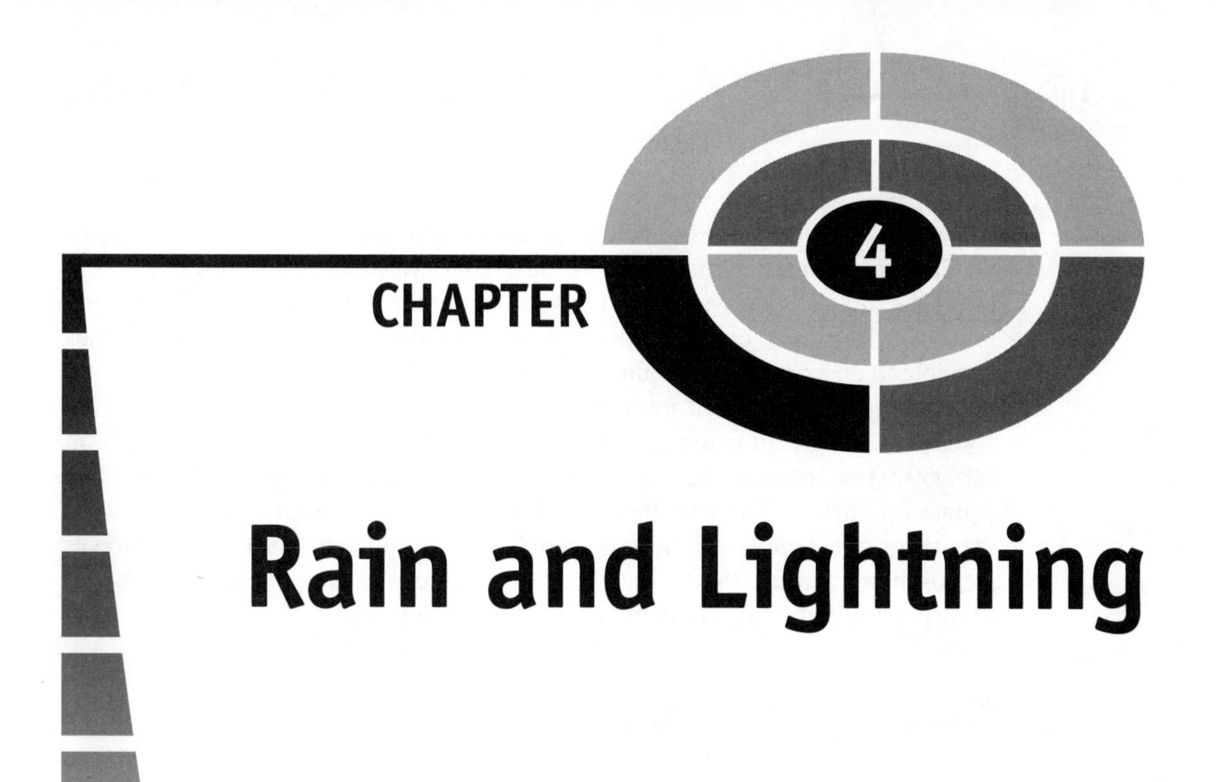

Rain and Lightning

Showers and thundershowers occur throughout the world. The rainfall they produce is often beneficial. In some cases, however, rain showers develop into destructive storms, or rain falls when there is already too much water.

Thunderstorm Formation and Evolution

Thunderstorms are always associated with cumulonimbus clouds. Such clouds can form all by themselves, or they can form in groups. When several thunderstorms band together and move as a unit, the combination is called a *squall line.* Squall lines are found in tropical cyclones, as well as in temperate low-pressure systems.

INITIAL STAGES

In the early stages of formation, a cumulonimbus cloud appears as an oversized cumulus cloud. The top is rounded, and little or no rain falls. Warm ground-level air ascends, pushing into the cooler air above and resulting in the formation of water droplets, making the cloud visible. As the warm air rises through the center of the cloud, more warm air flows inward at the cloud base, feeding energy into the system. The cloud rapidly grows in breadth and height. It is sometimes possible to observe a cumulonimbus cloud expand in its early stages, merely by watching it for a couple of minutes.

Eventually, precipitation begins. The cloud top can reach an altitude of 12 km (40,000 ft) or more above sea level, and its rain area may cover hundreds of square kilometers. The cloud edge contrasts with the surrounding sky. Finally, the updrafts encounter high-speed winds in the upper troposphere and lower stratosphere, and the top of the cloud is sheared off. This results in the well-known anvil shape, the signature of a fully developed thunderstorm.

TROPICAL AND SUBTROPICAL THUNDERSHOWERS

Thundershowers are an almost daily occurrence in rain forests near the equator. During the midmorning hours, the sun, rising high in the sky, heats the ground. This heat is conducted to the air at low levels, and the warm, moist air rises. By early afternoon, large cumulus clouds have formed. These clouds continue to grow, and rain begins to fall. The rain can become heavy. Gusty winds and lightning may accompany the shower.

Daily thundershowers are not unique to rain forests. Such cumulonimbus formation is common in the subtropics during the summer months. Florida and the Gulf Coast region of the United States provide good examples.

In a tropical or summertime subtropical thundershower, the clouds show little movement. The effect is local; it might be raining hard in one place, while the sun shines a few kilometers away. Such showers are relatively short lived. Once their rains have cooled the earth beneath, the source of heat energy is spent, and the showers dissipate. By early evening, the sky clears. The following day, the cycle is repeated.

Sometimes an advancing mass of cool or cold air pushes into the subtropics, intensifying the daily thermal showers. This situation occurs frequently in the spring, resulting in fast-moving, strong thunderstorms. High winds are characteristic; gusts can exceed 70 kt (80 mi/h). Small tornadoes are also common. Such storms are similar to squall lines that rake the Midwest in the summer. Although severe storms of this variety are most likely to take place during the afternoon hours, they sometimes occur in the late morning or early evening.

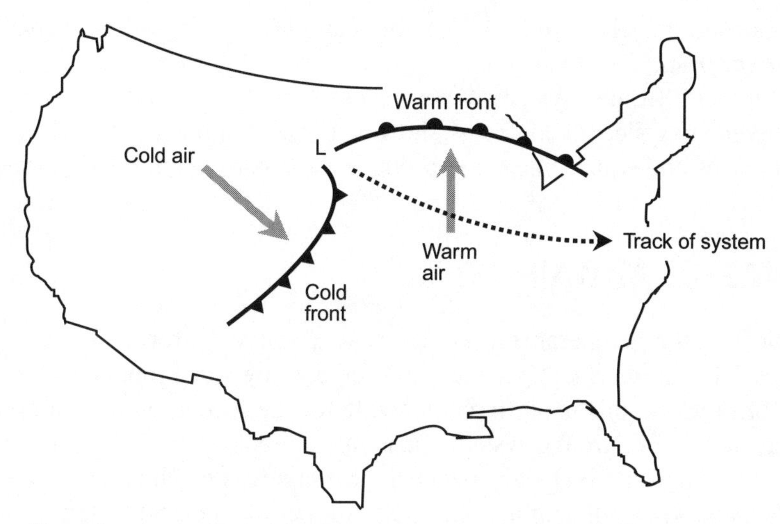

Fig. 4-1. A northern-hemisphere temperate low-pressure system over land, showing fronts, air flow, and the system track.

FRONTAL SYSTEMS

In the northern hemisphere, warm air flows from south to north along the leading sides of low-pressure systems, and cool or cold air streams from the north and west on the trailing sides. This results in a pattern similar to that shown by Fig. 4-1. In the southern hemisphere, warm air moves from north to south along the leading side of a clockwise-rotating low-pressure system; cold air blows from the southwest or west along the trailing side. Poleward-moving warm air produces clouds and showers near and ahead of its forward boundary, which is called a *warm front.* The forward boundary of the mass of cool or cold air, known as a *cold front,* is characterized by a well-defined band of clouds ahead of the frontal boundary and little or no cloudiness behind, producing a characteristic comma shape in satellite images.

In a temperate low, the cold front typically advances faster than the warm front. Because of this, part of the cold front eventually catches up with the warm front. The result is an *occluded front.* The *occlusion* is the region where the cold front catches and undercuts the warm front.

Sometimes a cold front trailing a low-pressure system, or a warm front ahead of one, loses its forward momentum and stalls over one area, producing prolonged rain. In this situation, foul weather can plague a region for days. This is a *stationary front.*

All types of frontal systems—warm, cold, occluded, and stationary—can cause severe weather. Cold fronts, however, are responsible for most heavy thunderstorms and squall lines, and produce the most violent wind storms.

AIR MASSES AT WAR

In stable air, the temperature decreases at a fairly uniform rate as altitude increases. This holds true for several kilometers upward. Frontal systems upset the uniformity, and this is partly responsible for the formation of rain clouds.

In the vicinity of a frontal system, the temperature drops at first with increasing altitude, but when the boundary between air masses is reached, the temperature abruptly increases. As the altitude increases further, the normal pattern resumes.

Fig. 4-2 is a vertical-slice diagram of a warm front. The thermal boundary is shown by the heavy curve. The warm air, because it is lighter than the cold air, tends to rise. As the front advances, the warm air flows over the top of the cold air mass, resulting in a *thermal inversion* at an altitude that depends on the horizontal distance from the front itself (the point at which the boundary between the air masses meets the surface).

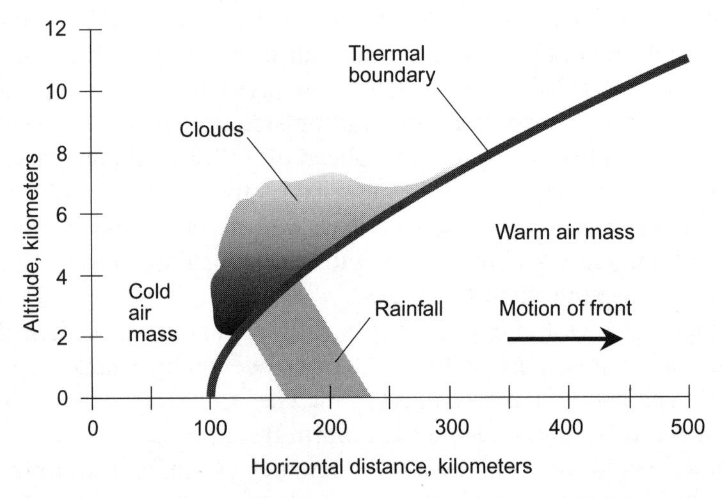

Fig. 4-2. Vertical-slice diagram of a warm front associated with temperate-zone low-pressure systems of spring and summer. The vertical scale is exaggerated.

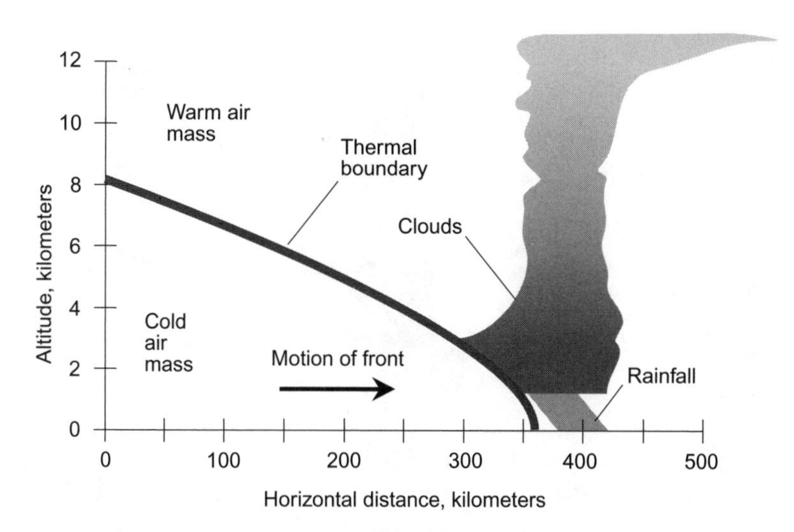

Fig. 4-3. Vertical-slice diagram of a cold front associated with temperate-zone low-pressure systems of spring and summer. The vertical scale is exaggerated.

The inversion in advance of a warm front can sometimes extend for hundreds of kilometers. Because a warm front usually moves at a sluggish pace—a few kilometers per hour—the cloudiness and rain can last for two or three days in some places. Warm fronts in the spring and summer sometimes produce severe weather, but usually they cause only moderate showers and thundershowers.

A cold front is more likely to give rise to severe thunderstorms. Figure 4-3 is a vertical-slice diagram of a cold front. Because cold air is more dense than warm air, the cold air mass pushes underneath the warm air mass. The leading edge of the front is well defined, and it advances rapidly along the surface. The temperature difference can be more than 10º C (18º F) at ground level at points separated by only a few kilometers.

If a cold front encounters irregular terrain, or if the front is moving fast, the leading edge of the cold air mass may "tumble over itself." This is because the air moves a little faster at an altitude of a few thousand meters than it does at the surface. If this happens, warm air pockets are trapped under the leading part of the cold air mass, creating powerful updrafts and turbulence as the warm air rises. Such conditions can produce large hail.

If an occluded front (Fig. 4-4) develops in a low-pressure system, severe weather can occur in the vicinity of the occlusion. The advancing cold air mass behind the front has a lower temperature than the air ahead of it, but the difference is not as large as it is when a cold front pushes rapidly into a mass of warm, moist air.

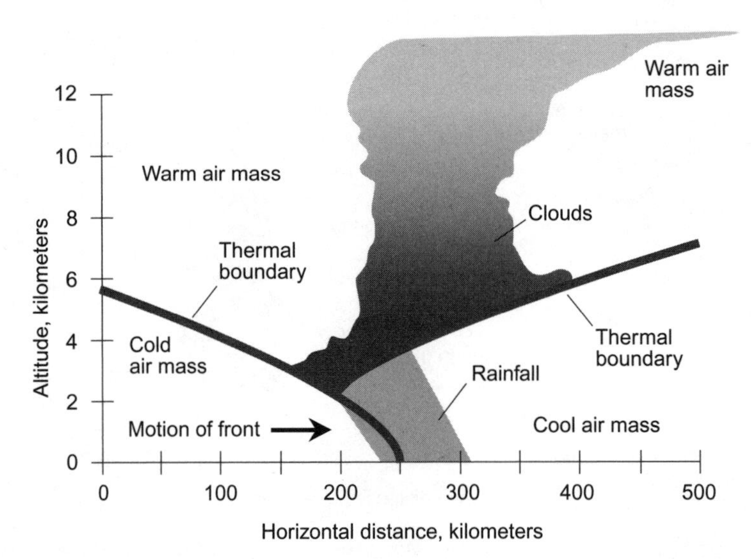

Fig. 4-4. Vertical-slice diagram of an occluded front associated with temperate-zone low-pressure systems of spring and summer. The vertical scale is exaggerated.

PROBLEM 4-1

What happens to a low-pressure system after an occlusion has occurred?

SOLUTION 4-1

A new warm front and cold front can form, as the winds around the center continue to pull warm air from the tropics and cold air from the poleward side of the system.

PROBLEM 4-2

What happens to the fronts in a low-pressure system as it moves away from land and over the ocean?

SOLUTION 4-2

Warm, cold, and occluded fronts can still form, and can still exist, in low-pressure systems over the ocean. The temperature difference between the warm and cold air masses in an oceanic low-pressure system is usually smaller than the temperature difference between the warm and cold air masses in a low-pressure system over land.

A Hypothetical Severe Thunderstorm

Tropical air, streaming up from the Gulf of Mexico and the southeastern states in the spring and summer, presents the ideal scenario for severe thunderstorm development, in conjunction with a fast-moving, intense cold front.

THE WATCH

When a severe thunderstorm watch is issued, it means that conditions are favorable for the development of such storms. Severe thunderstorm watches and tornado watches are often given together, because severe thunderstorms are more likely than less intense showers or thundershowers to spawn tornadoes.

Geographically, severe weather watch areas are nearly always shaped like rectangles or parallelograms. Heavy thunderstorms tend to develop along the most intense parts of a cold front. It is common for two or more widely separated regions to receive severe thunderstorm watches at about the same time, especially if there are two or more low-pressure systems over a continent, or if a system contains an unusually long cold front.

Here is a hypothetical severe thunderstorm watch announcement: "Severe thunderstorms, possibly containing large hail and damaging winds, are likely until 9:00 P.M. Central Daylight Time (CDT). The greatest risk of severe thunderstorms and tornadoes is along, and 80 km (50 mi) to either side of, a line from 40 km (25 mi) north of Sioux Falls, South Dakota, to 80 km (50 mi) southwest of Hayward, Wisconsin." This defines the boundaries of the watch parallelogram.

Most people living within the watch area will not experience a severe thunderstorm. Some will have a mostly sunny day and a pleasant evening. It's even possible that no severe storms will occur at all. The following day will be cooler throughout the region, after the passage of the cold front.

THE WARNING

Suppose, in the above described scenario, that we live in Minneapolis. We therefore are within the watch area. The first reports of severe weather come from the western and southern portions of the watch zone at 4:00 P.M. CDT. Worthington, Minnesota, gets 5 cm (2 in) of rain in 30 minutes, along with marble-sized hail and wind gusts to 55 kt (63 mi/h). The storm quickly passes.

In Minneapolis, there is no immediate cause for concern. The storm at Worthington, although heavy, is far from life threatening for those who stay indoors, away from windows (which can be broken by high winds or wind-driven hail) and electrical appliances, bathtubs, shower stalls, swimming pools, and antenna installations (which are hazardous because of lightning).

At 5:00 P.M., we hear the announcement that Minneapolis and its suburbs are under a severe thunderstorm warning. "A line of heavy thunderstorms has been sighted. Movement is toward the east–northeast at 26 kt (30 mi/h)." A group of thunderstorms, called a *multicell storm,* is approaching. The warning is effective until 6:30 P.M. for Hennepin County. A similar warning for Ramsey County will soon follow. Severe thunderstorm warnings are typically issued for much smaller areas than that of the associated watch. It is possible that severe thunderstorm warnings may have to be issued for localities not within the original watch rectangle.

SEVERE OR NOT?

Thunderstorms frequently develop as isolated cells, in which case they are called *single-cell storms.* Groups of two or three thunderstorms compose multicell storms; this is the type of system now bearing down on us. The heaviest thunderstorms develop internal vortices. Deep within the main cloud, a clockwise, or *anticyclonic,* vortex can form in the lefthand portion of the storm, and a counterclockwise, or *cyclonic,* vortex can form in the righthand half (as the cell is viewed from the rear). In the center of the thunderstorm, updrafts occur. These updrafts can reach speeds of 150 kt (170 mi/h). The double-vortex structure is illustrated in Fig. 4-5. This type of thunderstorm is called a *supercell.* The cyclonic vortex is called a *mesocyclone.*

The internal circulation of a severe thunderstorm is revealed by means of *dual-Doppler radar.* This instrument shows the locations of rain masses within showers and thunderstorms, and also indicates the velocities (speeds and directions) of water droplets, thereby showing the internal wind circulation. With the aid of computer graphics and animation programs, severe thunderstorms, monitored by dual-Doppler radar, can be vividly portrayed in three dimensions. When a thunderstorm has developed as a supercell structure, severe weather is likely, and tornadoes are a particular danger.

Sometimes a supercell will split into two separate thunderstorms after it has traveled some distance. The clockwise spinning vortex tends to turn toward the left as the thunderstorm moves forward, and the counterclockwise vortex veers toward the right. After the two vortices have separated sufficiently, the supercell divides. If the process is repeated several times over a period of hours, a single

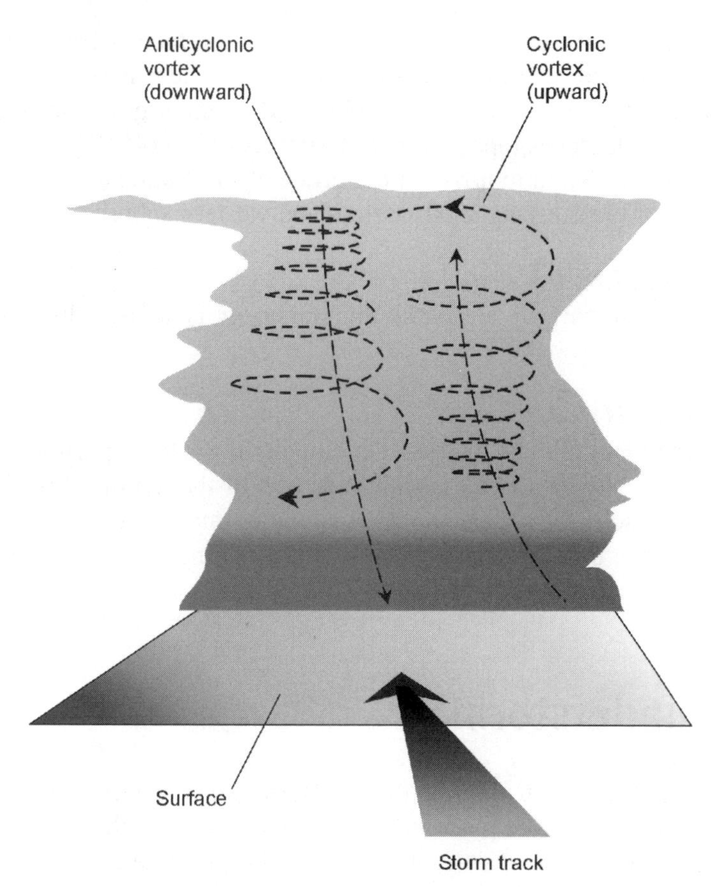

Fig. 4-5. The internal air flow of a typical supercell. Two vortices can form, one anticyclonic and the other cyclonic. Updrafts sometimes blow at 150 kt (170 mi/hr).

storm can become a multicell complex, and a group of thunderstorms can become a squall line.

A typical thundershower lasts for a few hours. Supercells last longer, up to half a day. The multicell and squall-line types of severe thunderstorms can persist for even longer, and are more likely than the single-cell storm to become severe.

Showers and thunderstorms form more often during the afternoon and evening hours than at other times of day. This can be explained by the fact that the most thermal energy (heat) is present during the afternoon and evening. The sun heats up the ground, and the warm earth imparts heat to the lower atmosphere, intensifying the contrast between the air temperatures on either side of a moving cold front.

In our hypothetical Minneapolis scenario, all of the factors for the formation of a severe storm are present, including a hot, sunny afternoon with plenty of ground heating. By 5:30 P.M., the sun is obscured by the tops of the cumulonimbus cloud bank, which has topped out at more than 14 km (46,000 ft). Reports of hail and high winds begin to arrive at the office of the National Weather Service from residents of the outlying southwestern and western suburbs.

PROBLEM 4-3

In which part of a severe thunderstorm is a tornado most likely to form?

SOLUTION 4-3

Tornadoes are most likely to develop in the cyclonic portion of a supercell. This tends to be the right-rear part of the storm. This is why tornadoes so often appear to follow along to one side of a supercell thunderstorm, and can strike when it is not raining and the weather seems to have calmed.

Effects of a Thunderstorm

The specific events in an intense thunderstorm—heavy rain, hail, damaging winds, and tornadoes—take place in different parts of a single cell. In multicell storms or squall lines, the chance of any particular place getting hit by multiple effects are greater than is the case with a single storm.

HEAVY RAIN

The heaviest rains tend to occur beneath the anticyclonic vortex in a heavy thunderstorm. The vortex develops a downdraft, which blows the rain down toward the ground. The result can be a blinding torrent, reducing visibility to near zero and producing spectacular amounts of rain in short periods of time.

HAIL

In a mesocyclone, powerful updrafts carry water droplets to great altitude. The cold high-level air causes the raindrops to freeze. The ice pellets fall back as they are scattered out of the updraft zone by high-altitude winds. Some of the ice pellets melt in the warmer, low-level air and fall to the ground as rain. Others are pulled back into the central updraft before they get a chance to melt. Water droplets condense on these pellets, and then freeze to enlarge the pellets' size as they are tossed upward again to high altitudes. A single pellet can be carried up and down repeatedly until it grows to such a size that, once it finally does defeat the updrafts and get to the ground, it arrives still frozen, as a hail stone.

Most of the hail in a severe thunderstorm falls near the center of the cell. The size of the hail stones depends on the strength of the updrafts in the storm; the more powerful the air currents, the more times a hail stone can be carried upward to acquire a new coating of ice. Hail stones are usually the size of peas or small cherries. Occasionally, golf ball-sized hail forms. In rare instances, hail stones become as large as baseballs or even grapefruit. Such large chunks of ice, falling from an altitude of several kilometers, can ruin farm crops, injure or kill livestock, and cause serious damage to property, particularly shingled roofs.

The amount of hail that falls, and the size of the stones, varies greatly within a single storm. Baseball-sized hail might fall in one part of a town, while little or no hail is reported in other places. Hail can sometimes accumulate to a depth of several centimeters, and if high winds are present, drifting can occur.

Hail storms in the northern hemisphere are most frequently observed during the months of April through September, with a peak around June and July. When hail falls in the late fall or winter, the stones are usually so small that it is difficult to say whether the precipitation is hail or sleet.

WIND

Damaging straight-line (nontornadic) winds in a severe thunderstorm can occur for various reasons. A powerful straight-line thunderstorm wind is called a *derecho* (pronounced da-*ray*-cho).

The heavy rain in the lefthand part of a storm and the accompanying downdraft from the anticyclone produce a rush of cool air ahead of the storm. Anyone who lives in the Great Plains of the United States has observed this; it occurs even in ordinary showers. This is known as a *gust front,* and the wind can rise from near calm to 50 kt (58 mi/h) or more within seconds. As the rain begins, the wind speed usually decreases.

Within a severe storm, high winds can result from the two vortices. These winds have been known to gust to more than 100 kt (115 mi/h) at the surface in massive supercells. Near the lefthand periphery of a supercell, the clockwise twisting of the air acts in conjunction with the downdraft and the forward movement of the cell to produce strong, straight-line surface winds.

Another cause of high straight-line winds, the least understood but potentially the most destructive, is known as a *microburst.* It results from a small column of air forced violently down toward the surface, and can produce straight-line winds of such force that cars are flipped over, whole groves of trees blown down, and roofs removed from buildings. In July 1977, a storm in and around Sawyer county, Wisconsin, caused microbursts with localized gusts of well over 100 kt (115 mi/h).

As we watch our hypothetical thunderstorm approach Minneapolis, the calm, sultry air is suddenly disturbed by a cool, westerly breeze. The breeze rapidly increases to a gale, and loose papers and other light objects are picked up and carried along. Dust, blown aloft, gives the sky a menacing hue.

LIGHTNING

The most dangerous aspect of a thunderstorm, from the standpoint of risk to human life, is lightning. It can occur anywhere within a thunderstorm, but is most common around the periphery of the cloud base. The amount of lightning in a thunderstorm is correlated with the severity. We will examine lightning in detail later in this chapter.

TORNADOES

Tornadoes rank just behind lightning in terms of deadly effects. They almost always occur in the cyclonic part of a supercell. Rarely, they form in the anti-cyclonic portion. In Chapter 5, we will see how and why tornadoes form, what they can do, and what we can do to protect ourselves and our property from them.

A SUPERCELL STRIKES

As we reach shelter, the rising wind slams the door back against the siding. It is not easy to pull a door—even a light screen door—shut against this gale. But finally it is latched, and we push the solid inner door closed behind it. It's a good thing we're inside. A large tree limb comes tumbling down into the yard. Foliage begins to litter the streets. The rain comes: a sprinkle at first, then sideways-moving sheets. It pounds with fury against the north and west windows of the house. As we hurry down the steps to the cellar, the electricity fails. For 30 minutes the winds rage. A portable radio tells us of widespread wind damage.

In the extreme southern suburbs, a tornado has been sighted, but it has not touched down. Close to downtown, tennis-ball-sized hail has been reported by unofficial observers.

The sounds of the wind and rain subside, and we return from the cellar to survey the damage. Many trees have been uprooted or snapped at the trunk. The tar paper from our garage roof is lying in the middle of the street.

The meteorologist at the airport is on the radio. He says that they have clocked winds of 75 kt (about 86 mi/h). The areas nearer downtown, including our neighborhood, have received a more severe blow. In some neighborhoods, hail stones have completely covered the grass, streets, and sidewalks. The tornado stayed above the surface, and no damage is reported from it. Power outages are widespread. Live wires dangle in puddles and hang, sparking, from utility poles.

STAYING SAFE

When a severe thunderstorm watch is issued for your area, tune the radio to a station that you like, and keep it playing in the background, continuously if possible. If you cannot listen continuously, then monitor the news broadcasts, usually given on the hour and sometimes also on the half hour. If time permits, get your car or other motor vehicle to a sheltered place to protect it against damage from hail or flying debris. Pick up loose yard tools, garbage cans, and other objects that could be thrown by high winds. If you have a boat on a trailer, put it in a sheltered place if possible. Otherwise tie it down with aircraft cable. (This is a good routine practice, anyway.) Stay at home unless you absolutely must travel.

The "little old ladies" who run around unplugging appliances before a storm are not stupid! Lightning can ruin electrical appliances and electronic devices if it strikes the power lines nearby. If you have any kind of radio or television apparatus that is connected to an outdoor antenna, disconnect the antenna feed line at the equipment before lightning starts to occur in your area. (If the storm has already begun, it's too late. In that case, stay away from the appliances!)

A severe thunderstorm can strike with amazing rapidity. The wind can rise from calm to gale force within seconds. If you have a basement, go there (unless it is prone to fill with water). Stay in the corner away from the wind, or in the center, or in a room without windows. Glass can break from the force of the wind or because of flying debris. Also stay away from electrical appliances, and avoid any kind of radio, television, or hard-wired telephone equipment. Stay out of elevators!

In some areas, heavy rains can produce *flash flooding*. If such an event is imminent, you will be informed by local authorities. If they recommend that you evacuate, do so immediately.

If the electricity goes out, switch off as many lamps and appliances as possible, especially those that draw large amounts of current or that might be damaged by a voltage surge when power is restored. Air conditioners, radios, and television sets should be shut off or unplugged.

In the aftermath of a severe thunderstorm, there may be damage to trees, buildings, and utility lines. Fallen trees might block roads, and standing water can accumulate to depths of several feet. Resist the temptation to get out and survey the damage. Downed power lines can cause electrocution by means of conduction through puddles of water. Fires can occur as a result of lightning, because of sparking utility wires, or because of damage to fuel tanks. Report fires immediately to your local fire department. If there are any injuries to people, take appropriate first aid action and, if necessary, call an ambulance. Do not make unnecessary telephone calls, even on cell phones.

PROBLEM 4-4

Why is a flash flood so dangerous? After all, it's only water. If you're a good swimmer, why should you worry?

SOLUTION 4-4

The water can rise so suddenly that there is no time to get away from it if you are caught in a low-lying area. Water can come downstream (or down the street) in a series of large waves, attended by a current that moves along at several meters per second. The current can carry heavy or sharp objects along, injuring people who get caught in the water, no matter how well they can swim. A person caught bodily in a flash flood can be rammed into, or under, an obstruction after being pulled underwater.

PROBLEM 4-5

Isn't it an overreaction to rush to the basement when a heavy thunderstorm strikes?

SOLUTION 4-5

It's better to overreact than to be injured by flying glass or caught unawares by an unseen tornado. However, if you live in a flash-flood area, and especially if you know your basement can fill with water, it is better to stay on the main level and go to a room on the leeward side of the building (the side facing away from the wind, usually the east side).

The Anatomy of Lightning

Thundershowers and thunderstorms produce dangerous and frequent lightning. So do hurricanes and, occasionally, snowstorms or ice storms. In the United States, lightning kills and injures dozens of people every year.

THE ATMOSPHERIC CAPACITOR

The earth is a fairly good conductor of electricity. So is the upper part of the atmosphere known as the *ionosphere*. The air between these two conducting regions, in the troposphere (where most weather occurs) and in the stratosphere, is a poor conductor of electric current. When a poor conductor is sandwiched in between two layers having better conductivity, the result is a *capacitor*.

A capacitor has the ability to store an *electrostatic charge*. This produces a potential difference (voltage) between the conductors of the capacitor. The amount of charge that a capacitor can hold is proportional to the surface area of the two conducting surfaces or regions. The *earth-ionosphere capacitor* is shaped like one huge sphere inside the other. It is nearly 13,000 km (8000 mi) in diameter (just a little larger than the diameter of the earth), resulting in a huge surface area between two spherical "plates" whose spacing is small compared with their diameters. There is a high voltage between the surface and the ionosphere (Fig. 4-6), and this gives rise to a constant *electric field* in the troposphere and stratosphere.

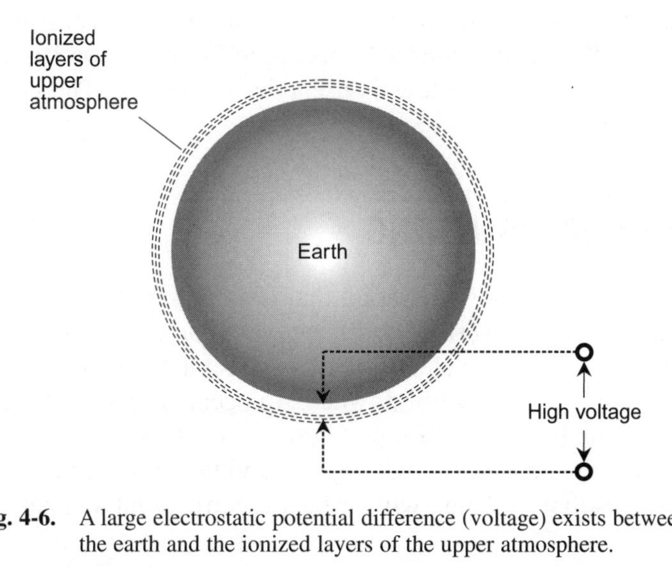

Fig. 4-6. A large electrostatic potential difference (voltage) exists between the earth and the ionized layers of the upper atmosphere.

The troposphere and stratosphere of the earth compose a good electrical insulator, but it is not perfect. Vertical air currents, and regions of high moisture content, produce *channels* of higher conductivity than that of the surrounding air. Cumulonimbus clouds, which sometimes reach to the top of the troposphere, present paths of better ground-to-ionosphere conductivity than stable, dry air. The average effective resistance between the ground and the ionosphere is about 200 *ohms,* roughly the same resistance as a 75-watt bulb in a household electrical circuit. In and near thunderstorms, the resistance is considerably less than this. Cumulonimbus clouds thus present an attractive environment for electrical discharges to occur. Some discharge occurs slowly and constantly as "bleedoff" throughout the world, but a fast discharge—an *electrical arc*—happens frequently. These arcs are lightning strokes. Numerous thunderstorms are in progress on our planet at any given moment, helping to limit the charge between the ground and the upper atmosphere.

The atmospheric capacitor maintains a constant charge of about 300,000 volts, which is typical of high-tension utility lines. The average current that flows across the atmospheric capacitor is about 1500 amperes. Hence our atmosphere is constantly dissipating about 450 megawatts of power—the equivalent electrical usage of a medium-sized city.

Most of the current flow between the upper and lower atmosphere takes place within cumulonimbus clouds. They concentrate the charge, increasing the local voltage. They also tend to reduce the space between opposite electric poles in the atmospheric capacitor, creating a particularly attractive place for discharge. A typical thundershower discharges about 2 amperes of current, averaged over time. At any given moment, there are 700 or 800 thundershowers in progress on our planet. A current of 2 amperes per thunderstorm may seem small, but this current does not flow continuously. It occurs in brief, intense surges. A single lightning discharge lasts only a few thousandths of a second. Therefore, the peak current is extremely large.

FREQUENCY OF LIGHTNING

Numerous factors affect the likelihood that a lightning stroke will directly hit any particular point on the surface within a given span of time. Lightning almost always occurs in or near large clouds, but it is possible for a "bolt from the blue," which seems to strike from out of nowhere, to occur. The size of a cumulonimbus cloud base, the height of the cloud base above the ground, the intensity of the precipitation, the presence of tall objects, and the conductivity of the soil all affect the chances that lightning will strike a specific point at a particular time.

In the United States, thunderstorms occur most frequently in central Florida. Many thunderstorms also take place in the Rocky Mountain states and in the Midwest. Somewhat fewer lightning strikes are observed in the Northeast. Very few strikes occur in the northwestern states, in spite of the fact that much of this part of the country receives abundant rainfall.

Warmth and moisture are both very important to the formation of thundershowers. Strong vertical updrafts and downdrafts also contribute. Most of the charge concentration in the atmosphere takes place in the lower latitudes, particularly in the tropics. The least charge concentration is seen near the poles.

THE CHARGE IN A CLOUD

In a storm cloud, the temperature decreases rapidly with increasing altitude. There are strong updrafts and downdrafts present in the system, and this can generate a large separation of electric charge as well as severe weather. Positive charge accumulates near the top of a cumulonimbus cloud, and negative charge accumulates near the base of the cloud, as shown at A in Fig. 4-7. The resulting pair of charge centers is called a *cell.* It actually is an electric cell, with a potential difference that can reach millions of volts.

In some cases, a negative charge accumulates at the top of a tall cumulonimbus cloud, above the positive layer. This forms a second electrical cell, at a higher altitude than the first. There may also be differences in charge in a horizontal direction. One part of a cloud, at a given altitude, may have a different voltage than another part of the same cloud at the same altitude. A single thundershower can have several cells within itself, and thus several possibilities for lightning discharges.

Lightning within a single shower, or between nearby clouds, is the most frequent type of lightning. It never hits the ground, although we see the flash and hear the thunder. At night, it is possible to see cloud-to-cloud lightning from more than 100 km (about 60 mi) away. If the flashes occur below the visible horizon and are dispersed by high-altitude clouds such as cirrus, the result is known as *sheet lightning.*

CLOUD TO GROUND

A common type of lightning discharge between a cloud and the earth occurs as a flow of electrons from the base of a thundershower to the surface. The base of the cloud carries a negative charge, inducing a localized positive voltage on the

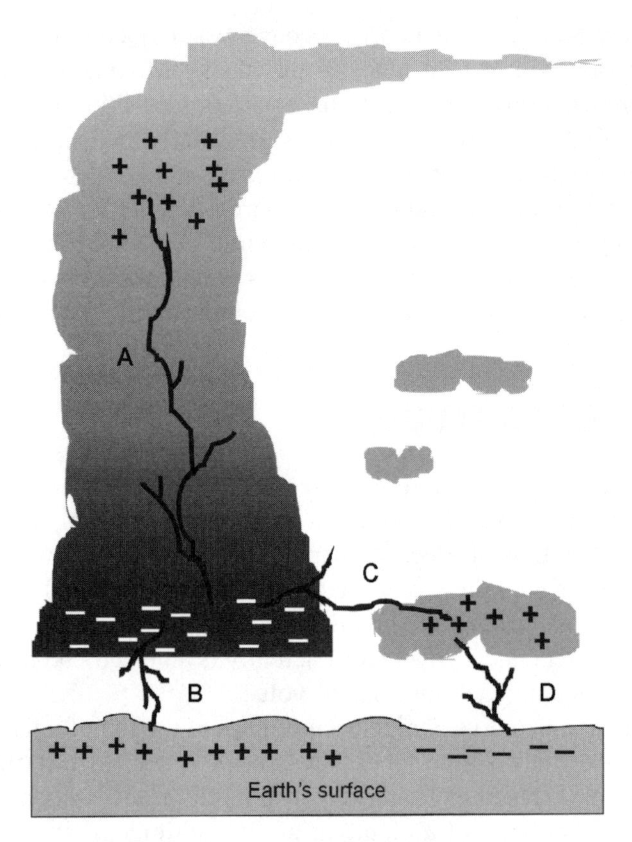

Fig. 4-7. Four types of lightning stroke: intracloud (A), cloud-to-ground (B), intercloud (C), and ground-to-cloud (D).

surface beneath (at B in Fig. 4-7). The charge concentration on the ground can get so great in some locations that people's hair stands on end.

When the voltage builds up sufficiently and the air can no longer maintain effective insulation, a streamer of electrons, carrying negative charge, begins to probe outward and downward from the base of the cloud. This is called a *stepped leader.* The electrons seek the path of lowest resistance to the ground, in the same way that a river finds the easiest route from a mountain to the sea. There are many dead-end paths that have good conductivity for a short distance, but then terminate. As the stepped leader approaches the area of positive charge at the surface, the *air dielectric* breaks down completely.

When the stepped leader is about 100 to 200 m (330 to 660 ft) from the ground or the object about to be stricken, a flashover of positive charge takes place. The flashover meets the downward moving leader a short distance above the ground. This point is called the *point of strike.* The electrical circuit is complete; the stepped leader has generated an ionized, highly conductive path from the cloud base to the ground. A massive discharge immediately follows, beginning at the ground and progressing upward to the cloud. (Although the individual electrons travel downward, the discharge as a whole takes place in an upward direction.) This is the *return stroke,* and it is responsible for the visible flash. The diameter of a typical return stroke is only a few centimeters, but it is as hot as the surface of the sun. Fires can be started within the few milliseconds (ms) during which the current flows.

After the return stroke, another leader may move downward from the cloud base. Because an ionized, conducting path has already been established by the first stroke, this leader moves much faster than the original leader, and is called a *dart leader.* As the dart leader reaches the point of strike, there is another return stroke. A third leader may follow, causing a third return stroke. The whole process can be repeated several times.

A complete lightning stroke lasts about 20 ms if there is only one return stroke. The flash can last as long as 500 ms if 10 or 15 return strokes occur. The greater the number of return strokes, the brighter is the flash and the louder is the resulting thunder. A *multiple stroke* presents a much greater fire hazard than a single stroke, and is also more deadly.

OTHER TYPES OF DISCHARGE

Intracloud lightning (shown at A in Fig. 4-7) and *intercloud lightning* (shown at C) discharges occur often in and near thunderstorms. In a multicell storm or a squall line, zones of maximum positive and negative charge can arise at various altitudes, creating a multiplicity of possible discharge paths. When lightning occurs inside a distant cloud at night, the whole towering storm is illuminated in an eerie and spectacular way. The lightning does not damage objects on the ground, except perhaps indirectly as a result of the strong *electromagnetic* (EM) *field* that all lightning bolts generate.

Although the base of a cloud usually develops a negative charge, and the ground beneath a cloud accumulates localized positive charge, the situation is sometimes reversed. This causes *ground-to-cloud lightning* (shown at D in Fig. 4-7). The electrons flow upward rather than downward. Ground-to-cloud lightning takes place in exactly the same way as a cloud-to-ground discharge, but upside down. The stepped leader starts from the surface or the object to be stricken and advances upward until it nears the center of positive charge in the cloud. The positive flashover meets the stepped leader at a distance of about 300 to 400 m (1000 to 1300 ft) from the center of positive charge; this is the point of strike. The return stroke transfers the positive charge downward (although the electrons actually flow upward). Ground-to-cloud lightning is usually less destructive to a stricken object than cloud-to-ground lightning.

In some situations, there exists a large potential difference between two points, but it is not enough to produce a full discharge. The stepped leader fails to complete the circuit, and so it simply dies out. This is called an *air discharge.* The air discharge causes little or no thunder, and there is rarely any damage to objects on the ground.

The most intense form of lightning discharge takes place when the positive voltage at the top of a cloud builds up to such magnitude that a flashover occurs all the way from the surface. This is known as a *superbolt,* because it carries more current than an ordinary lightning stroke. A superbolt can strike the ground at a considerable distance from a cumulonimbus cloud. It causes a great clap of thunder, and can cause extensive damage because of the large amount of electric current that flows.

Some odd forms of lightning are sometimes seen. *Ball lightning* is one such strange phenomenon. People have reported seeing spheres of fire that persist for several seconds. The spheres range in diameter from a centimeter or two to more than a meter. A "ball of fire" may be seen to roll down the hall from the living room to the kitchen, and then disappear in a puff of smoke. Ball lightning can enter buildings through electric utility or telephone wires.

Saint Elmo's fire is a slow discharge that can be eerie but is rarely destructive. When a thundershower is nearby or overhead, metal objects such as radio antennas, lightning rods, and sailboat masts sometimes acquire a phosphorescent halo. Saint Elmo's fire is often observed on airplane wingtips. Some airplanes have pointed rods on the wingtips and tail structure to encourage this slow discharge. The idea is that slowly bleeding off an accumulation of electrical charge reduces the risk that the aircraft will be hit by a major stroke.

THUNDER

When a return stroke occurs in a lightning bolt, the air is heated almost instantaneously to a temperature of several thousand degrees Celsius. The heating causes the air to expand with great force, generating a shock wave that we hear as thunder. The sound can be heard for varying distances, depending on the wind direction and speed, the presence or absence of favorable sound propagation conditions, and the amount of background noise. Usually, thunder is audible up to about 8 km (5 mi) from the point of strike.

If the lightning flash can be seen, the thunder is heard somewhat later. Sound travels through air at a speed of approximately 335 m/s, which is roughly 1 km per 3 s or 1 mi per 5 s. You can determine the distance to a lightning flash by counting the number of seconds it takes for the thunder to arrive. Divide the number of seconds by 3 to get the distance in kilometers, or by 5 to get the distance in miles.

The rumbling, or booming, noise of thunder puzzles some people. A lightning flash has short duration, but thunder seems to last for several seconds. This occurs for two reasons: echoes and propagation delays. In hilly or mountainous terrain, or in cities with many tall buildings, the acoustic noise from a lightning stroke gets a chance to bounce around. We hear not only the original thunder, but its echoes. This creates a prolonged rumble. In flat, open country, or on a lake or at sea, there are no objects to cause the echoes, and the rumbling is less pronounced. Propagation delays also contribute to the rumbling effect. Suppose you stand 1 km from a stricken object, and the lightning flash occurs vertically from a cloud base 1 km high. You are about 400 m closer to the bottom of the lightning bolt than you are to the top, and the sound is therefore spread out over a time interval of slightly more than 1 second.

THE ELECTROMAGNETIC PULSE

In a lightning stroke, electrons are rapidly accelerated as they jump the gap between poles of opposite charge. Whenever electrons are accelerated, EM fields are produced. A radio broadcast transmitter works according to this principle; electrons are made to accelerate back and forth in the antenna at a precise and constant frequency. In a bolt of lightning, electrons are accelerated in a haphazard way, and this produces *radio noise* over a wide range of frequencies. The radio "signals" from thunderstorms are known as *sferics.* Sometimes the sound they produce in a radio receiver is called "static," although technically this is a misnomer.

Sferics travel for long distances at low radio frequencies, and for progressively shorter distances as the frequency becomes higher. A fairly distant group of thundershowers can cause interference to low-frequency radio communications. Low-frequency radio direction-finding (RDF) apparatus can actually be used to locate regions of intense thunderstorm activity. As thundershowers approach, the sferics become more frequent and intense, and they are observed at increasingly higher radio frequencies. A general coverage shortwave radio receiver can be used to demonstrate this effect. At first, sferics are heard only on vacant channels near the bottom of the standard AM (amplitude modulation) broadcast band at 535 to 1605 kilohertz (kHz). Then the noise becomes discernible on occupied channels. Finally, the sferics can be heard at frequencies up to several megahertz (MHz), and an occasional burst of noise punctuates the airwaves even in the standard FM (frequency modulation) broadcast band at 88 to 108 MHz.

A thundershower can generate EM fields of much greater intensity than any human-made radio transmitter. Such a burst of energy, called an *electromagnetic pulse* (*EMP*), can induce large currents in utility wires, telephone lines, radio antennas, fences, and other metallic objects. These currents can damage electrical appliances and electronic devices, and can give rise to high voltage "spikes" that are dangerous to people and animals. This is why you should never use electrical appliances, telephones, or radio equipment during a thunderstorm.

PROBLEM 4-6

Can the EMP from an intercloud or intracloud lightning stroke cause damage to electronic equipment located on the ground?

SOLUTION 4-6

Yes. It is not necessary for lightning to strike the surface, or to directly hit anything at all, in order for the EMP to produce destructive currents in utility wires, radio antennas, or telephone lines.

Protection from Lightning

There are certain precautions you can take to minimize the hazard, both to yourself and to your property. The following measures have been recommended by engineers. However, there is no such thing as absolute protection from lightning.

LIGHTNING RODS

If you have ever taken a drive on a back road in the Great Plains, you have seen *lightning rods.* A lightning rod is a metal conductor, usually steel, erected verti-

cally on top of a building or house. The lightning rod is electrically connected to an earth ground by a heavy wire.

When a lightning rod is installed, the conductor running to the ground should be as heavy as possible, preferably American Wire Gauge (AWG) No. 4 or larger. Copper tubing is also good. The conductor should follow the most direct possible path to ground along the outside of the building or house. The grounding conductor for a lightning rod should never be run inside a building or near flammable materials. The ground connection should be made to a *ground rod,* preferably at least 2.5 m (about 8 ft) long, driven into the earth a couple of meters away from the foundation of the house. An enhanced ground connection can be obtained by using several ground rods connected together.

In theory, a lightning rod provides a "cone of protection" that has an apex angle of 45°. You can calculate the approximate minimum height that a lightning rod must have, in order to provide protection for your house or building. Measure the distance (call it d) between opposite corners of the gables, divide by 2, and then add 20%. This will give you the minimum height (call it h) for the tip of the lightning rod, as measured with respect to the plane in which the corners of the roof lie (Fig. 4-8). Thus:

$$\begin{aligned} h &= 1.2\ (d/2) \\ &= 0.6\ d \end{aligned}$$

The risk of a direct hit within a lightning rod "cone of protection" is reduced, as compared with the risk of a direct hit outside the cone, or the risk of a direct hit in a similar situation without a lightning rod. This reduces the chance that a thunderstorm will set a house or building on fire. However, even the best lightning rod cannot completely eliminate the possibility of a direct hit within its "cone of protection." Lightning rods should not be regarded as a way to eliminate all danger. If a lightning rod has been installed, all the other standard precautions should still be taken. *Warning: If you are interested in installing a lightning rod to protect your house, consult a professional engineer and contractor. Don't try to put it up yourself.*

PROTECTING APPLIANCES

If you have had one or more lightning rods installed, do not become overconfident. If lightning strikes the rod, or even if it strikes in the neighborhood, the resulting EMP can damage electrical appliances and electronic equipment in your house. The chances of damage can be reduced by unplugging computers,

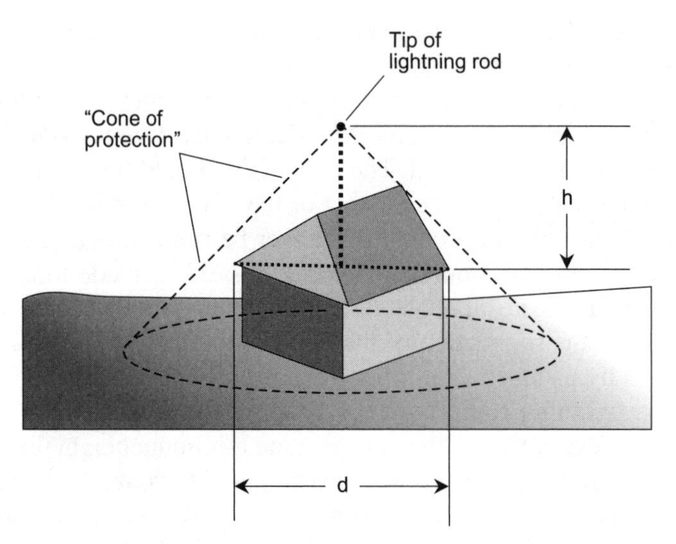

Fig. 4-8. A lightning rod can provide a "cone of protection" within which the probability of a direct hit is low.

amateur and shortwave radios, and other electronic devices that contain solid-state components. All radios and television sets should be disconnected from outdoor antennas. If your computer has a telephone modem, it should be disconnected by physically unplugging its cord.

PERSONAL SAFETY

Stay away from windows during a thunderstorm. Thunder can produce shock-waves strong enough to break large panes of glass. High winds and hail can break windows. Lightning can actually enter a house or building through a window. If lightning strikes near a window, tree branches or other debris can be thrown through the glass.

If you are caught outdoors in an electrical storm with no way of getting to shelter, you should move to a low place away from tall structures, and crouch down with your feet close together and your face toward the ground. A car or truck can provide protection from lightning, if you can get into one. Do not, however, stand near isolated tall objects such as flagpoles or trees! These objects "attract" lightning (meaning that lightning is most likely to hit them because they present a low-resistance discharge path). If you are near such an object when lightning strikes it, some of the discharge may also pass through you.

If you are caught in a boat during a heavy electrical storm, your situation is particularly dangerous, especially if your boat has a radio antenna or a tall mast. You should crouch in a position facing away from the antenna or mast, with your face down and your feet close together. (You might also want to cover your ears.) If the boat has an enclosed space, you should get inside it and stay away from metal objects and the boat's radio equipment.

All of the precautions mentioned here are intended to minimize personal risk. However, lightning has a capricious nature, and can be dangerous or deadly even when people think they are entirely safe from it. There are no guarantees. The Native Americans of the Great Plains knew this, and had a healthy fear of lightning.

BENEFICIAL EFFECTS

Lightning is not all bad. It generates nitrogen compounds that aid in the fertilization of the soil. Lightning limits the charge that builds up in the atmospheric capacitor. The lightning strokes that take place in thunderstorms are mere sparks compared to the discharge that would ultimately take place if the earth–ionosphere voltage were to rise to a high enough level.

Some scientists think that lightning was a key event in the development of life on our planet. It has been suggested that the first organic substances, the *amino acids,* were formed from inorganic compounds as a result of chemical reactions requiring electrical discharge of the sort that occurs in a lightning stroke.

PROBLEM 4-7

Suppose your house measures 16 m diagonally, between corners the roof. How high above the plane containing the corners of the roof should a lightning rod extend, in order to provide reasonable protection?

SOLUTION 4-7

Use the formula given above. Let $d = 16$ m. Then $h = 0.6$ d $= 9.6$ m. Remember that this is the minimum height for the tip of the lightning rod, as measured above the plane containing the corners of the roof. If that plane is, say, 3 m above the surface of the earth, then the height of the rod relative to the surface must be (9.6 + 3) m, or 12.6 m.

Quiz

This is an "open book" quiz. You may refer to the text in this chapter. A good score is 8 correct. Answers are in the back of the book.

1. How does a severe thunderstorm watch differ from a severe thunderstorm warning?
   (a) In a warning, storms are likely to occur over a wide region, but in a watch, storms already exist and are under observation.
   (b) In a warning, a storm has been detected and is expected to strike, but the issuance of a watch means only that conditions are favorable for severe thunderstorm development.
   (c) In a warning, the conditions are favorable for severe thunderstorm development, but the issuance of a watch means only that a cold front exists.
   (d) In a warning, an occluded front has actually formed, but in a watch, an occluded front is only a possibility.
2. A microburst can be responsible for
   (a) strong gusts of wind in a thunderstorm.
   (b) large hail in a tropical shower.
   (c) numerous tornadoes in otherwise tranquil weather.
   (d) lightning from a cloudless sky.
3. Statistically, the most lethal characteristic of a thunderstorm is
   (a) the derecho.
   (b) the lightning.
   (c) the hail.
   (d) the tornadoes.
4. An occluded front occurs in a low-pressure system when
   (a) a cold front overtakes a warm front.
   (b) a cold front stops moving.
   (c) a cold front reverses direction.
   (d) the system becomes anticyclonic.
5. The "static" you hear on a radio receiver when a thunderstorm approaches is caused by
   (a) downed or falling power lines.
   (b) tornadoes in the early stages of formation.
   (c) hail stones striking each other at high altitudes.
   (d) electromagnetic fields.

6. Which of the following statements about lightning is false?
   (a) A lightning stroke can induce a strong current surge in nearby utility wires and telephone lines.
   (b) A lightning stroke is accompanied by acceleration of electrons through the atmosphere.
   (c) A lightning stroke can never make a direct hit within the "cone of protection" provided by a properly installed lightning rod.
   (d) A lightning stroke can damage electronic equipment even if a direct hit does not occur.
7. Severe weather can accompany any type of front, but the most likely type to produce squall lines is the
   (a) warm front.
   (b) stationary front.
   (c) occluded front.
   (d) cold front.
8. The top of a typical thunderstorm cloud exists at an altitude of approximately
   (a) 1200 m above sea level.
   (b) 12,000 m above sea level.
   (c) 120,000 m above sea level.
   (d) 1,200,000 m above sea level.
9. Mesocyclones are characteristic of
   (a) large low-pressure systems.
   (b) tropical rain showers.
   (c) all low-pressure systems.
   (d) supercell thunderstorms.
10. Thundershowers in tropical rain forests occur
   (a) only in high-pressure systems.
   (b) only in conjunction with strong cold fronts.
   (c) almost every day.
   (d) mainly in the winter.

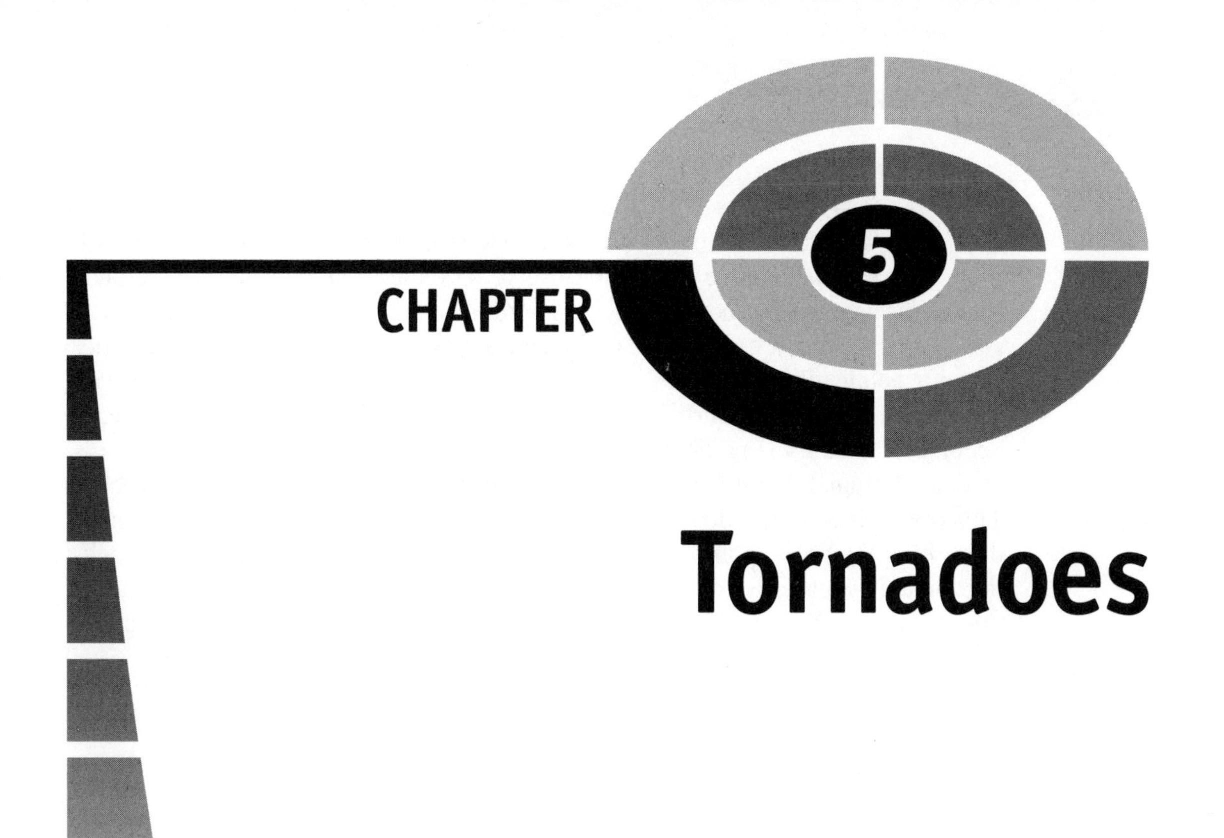

CHAPTER 5

Tornadoes

The onset of spring brings gentler weather to North America, Europe, and Asia than the months of December, January, and February. The massive cyclones of winter give way to smaller, less intense systems. But ironically, springtime also brings most tornadoes to the land masses, and particularly to the continental United States. The latitude, the geography of the coastlines, and the topography of the surface all conspire to make the central part of the "lower 48"—the Great Plains—the site of the most violent windstorms on earth.

Formation and Prediction

Large thunderstorms can, as we have seen, acquire a complex circulation. In a northern-hemisphere thunderstorm of this type, descending air rotates clockwise in the left-hand (usually northern or northwestern) part of the system, and rising air spins counterclockwise in the right-hand (usually southern or southwestern) part. Supercells develop a generally counterclockwise rotational circulation or mesocyclone. Mesocyclones are often associated with tornadoes.

A mesocyclone can grow stronger and tighter if the air pressure drops at its center. As the pressure continues to fall within the cyclonic vortex, the air spins.

This leads to a further drop in pressure, which causes the air to speed up still more. The effects become visible to observers on the ground because the clouds rotate. Air near the surface is drawn upward to feed the vortex. A vertical *funnel cloud,* shaped like a cone, rope, or cylinder, appears at the base of the "mother cloud" (technically called *mammatocumulus*). The twisting vortex works its way downward toward the surface. If the funnel cloud extends to the surface, it becomes a *tornado* (Fig. 5-1).

Once a funnel cloud has touched down, it kicks up clouds of dirt and debris. This often gives a tornado a dark hue. The cloud moves along the ground, sometimes hopping back up into the mother cloud for a few moments and then touching the earth again. Some tornadoes become low and wide; others become so thin as to appear thread-like. The diameter of a tornado can be as small as a few meters or as large as 1.5 to 2 km (about 1 to 1.25 mi).

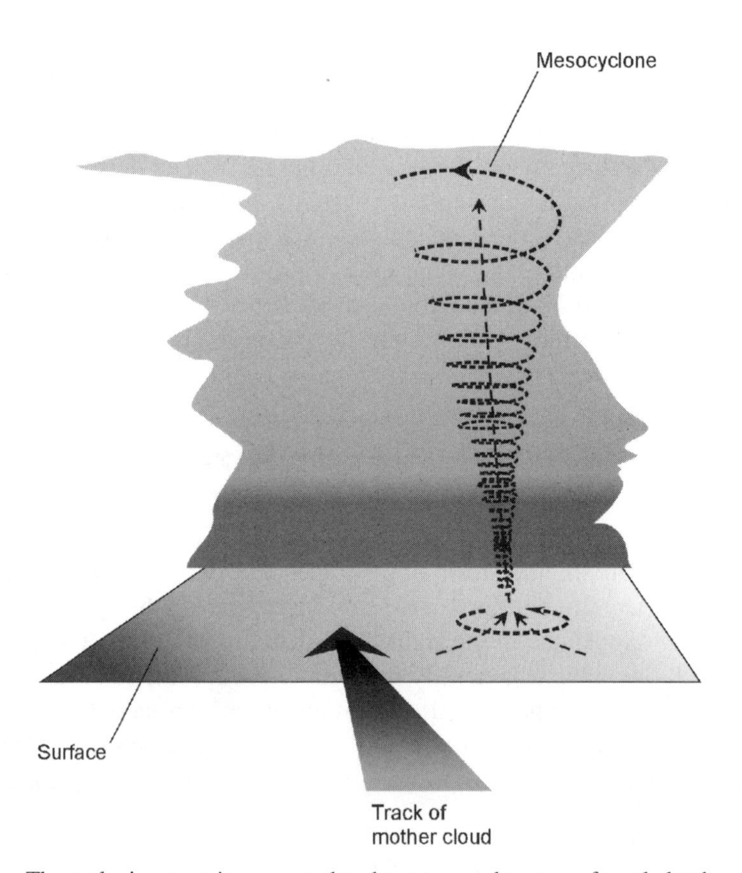

Fig. 5-1. The cyclonic vortex in a severe thunderstorm can become a funnel cloud or tornado.

As the tornado moves across the countryside, its general path coincides with that of the mother cloud, but the point of contact with the surface can be difficult or impossible to predict with absolute accuracy. Some tornadoes loop around after striking a particular point on the surface, and then hit that same point again. The funnel cloud may swing back and forth like the trunk of an elephant, or drill along in a straight line. After a period of a few minutes, most tornadoes begin to dissipate, although some last an hour or more. Depending on the population density of the stricken area, the damage can range from inconsequential to catastrophic.

TORNADO WATCH

Tornadoes are most common in the spring. A warm, muggy day in October is far less likely to bring a tornado in the northern hemisphere than a warm, muggy day in May. In the extreme southern United States, most tornadoes take place in March and April. In the middle of the country, April and May are the most common months. In the northern plains, May and June bring most of the tornadoes.

Tornadoes are rare west of the Rocky Mountains. They are also unusual in New England. The Plains states are most subject to tornadoes, but tornadoes can strike anywhere in the 48 continental states. They have even been seen in Alaska and Hawaii.

Assuming that you live in a tornado-susceptible region, and it is a spring or early summer day, you should be familiar with the type of weather that is associated with tornadoes. Some signs are a falling barometer, a southerly or southeasterly wind, and humid air, all occurring together. Tornadoes always take place in conjunction with thunderstorms, especially large or severe ones.

If you suspect that conditions are right for tornadoes, listen to the weather forecasts on local radio stations, or check a weather information Web site such as The Weather Channel (www.weather.com). Officials will issue a *tornado watch* for a specific geographic region if it appears that tornadoes are likely to form in that region. Tornado watches are often issued in conjunction with severe thunderstorm watches. A typical tornado watch reads or sounds like this: "The possibility of tornadoes exists until 6:00 P.M. along, and 100 km (60 mi) either side of, a line from 50 km (30 mi) southwest of Omaha, Nebraska, to 16 km (10 mi) north of Sioux Falls, South Dakota." Specific counties and towns are usually mentioned, because not everyone has a map and pencil handy. When the watch period expires, a new watch might be issued for a different area, the watch period might be extended, or the watch might be cancelled altogether.

The announcement of a tornado watch does not mean that a tornado has been sighted. Many, if not most, of the locations in a tornado watch area will not experience severe weather of any kind. It is possible that no tornadoes will occur anywhere within a watch area. But more tornadoes form per square kilometer in watch areas as compared with nonwatch areas. Once in awhile, a tornado occurs when no watch has been issued.

TORNADO WARNING

A *tornado warning* means that there is an immediate threat to a particular area. A tornado warning is usually issued for a much smaller region than a tornado watch. Tornado warnings are given whenever a funnel cloud is actually seen by official observers, or whenever radar indicates the presence of a mesocyclone in a thunderstorm.

A funnel cloud or tornado does not show up directly on a radar display, but the cyclonic vortex, deep within a heavy thunderstorm, produces a characteristic signature called a *hook-shaped echo*. Raindrops, spiraling upward with the whirling air, create the echo. The development of a strong mesocyclone, visible as a hook-shaped echo, often precedes a tornado. Radar can provide up to several minutes of warning prior to the formation of a funnel cloud.

In cities equipped with civil defense sirens, a tornado warning is announced by a continuous tone. The sirens are activated for 3 to 5 minutes, but the sirens are not within earshot of everyone. If there is heavy rain, a high wind, or hail in progress, some people will not hear the sirens blowing. If the tornado warning is issued in the middle of the night, some people will sleep through the faint wailing of distant sirens. Some municipalities have overcome these problems by installing numerous sirens.

For people who live outside of cities, weather monitoring radio receivers provide the best means by which tornado warnings can be made known. This type of radio, which is battery powered, stays silent until a tornado warning is issued. If a warning is put out, an actuating signal, transmitted by the weather officials in the vicinity, causes the unit to emit a loud tone or series of beeps. Anyone who lives in a tornado-prone region without adequate civil defense sirens should have one of these units. They are available from electronics stores.

As soon as a tornado warning is issued, a secure shelter should be found until the all-clear is broadcast over local radio and television stations.

RECOGNIZING A TORNADO

Not every ominous-looking cloud presents the danger of a tornado. People often mistake unusual, but benign, clouds for funnel clouds. High winds or moderate turbulence near a cloud base can make the base of a cloud irregular, and sometimes the ragged edges look like funnel clouds. There are certain easy-to-recognize signs, however, that indicate the presence of a tornado.

The first thing to look for is rotation at the base of a cloud. One part of a cloud may be moving toward the northeast, while another part drifts toward the southwest. The rotation is not necessarily rapid. In the northern hemisphere, the rotation is usually counterclockwise (as viewed from above) or clockwise (as viewed from beneath).

Another sign of a developing tornado is the appearance of obvious violent updrafts or downdrafts near the base of a cloud. Pieces of the cloud may be "torn off" and descend. Cloud fragments may mysteriously appear and shoot upward from the ground into the main body of the cumulonimbus. If a funnel cloud forms, it often appears smooth and well defined. There may be more than one of them, and as a group they usually revolve around each other. If a funnel cloud touches down, a cloud of dust (or water spray, if the funnel is over a lake or the ocean) usually appears at its base.

Not all tornadoes present themselves with fanfare. Some tornadoes are hidden within masses of low cloud. This type of tornado appears as a wide, boiling mass of darkness with no apparent shape. As it moves along, it seems to swallow everything in its path. These tornadoes are particularly dangerous because they can be mistaken for clouds of blowing dirt. Direct visual observation of tornadoes is not always possible. A funnel cloud can't be seen in rural areas at night. There might be a strong wind with heavy rain, which can reduce visibility to near zero.

Once a funnel cloud touches the ground, it produces a rumble or roar, the sound of which has been described as similar to that of a long freight train passing, a jet aircraft circling, or a huge stampede of cattle. In the open country, the sound of a tornado can sometimes be heard from several kilometers away.

RADIO AND TELEVISION TESTS

The interior of a funnel cloud contains almost continuous lightning. Lightning produces a familiar crackling sound on radio receivers. Lightning also causes a darkened television (TV) screen to light up momentarily, if that television set is connected to a set of "rabbit ears" or an old-fashioned outdoor antenna. Thus a radio receiver or television set can be used in some instances to detect a nearby funnel cloud.

Suppose a portable radio is tuned to a frequency in the standard amplitude-modulation (AM) broadcast band on which no station is heard. Normally, a moderate hiss or roar comes from the speaker or headset when the volume is turned up. In the vicinity of a thunderstorm, the hiss is accompanied by frequent popping and crackling noises because of numerous lightning strokes. But a funnel cloud, with its continuous discharges, creates an uninterrupted noise on the radio. The normal background hiss becomes much louder and may acquire a buzzing or whining aspect.

The so-called *TV tornado detection test* requires that a TV receiver be connected to a simple antenna, not to a cable or satellite dish. The antenna can be either indoors or outdoors, but it is best to use an indoor antenna because of the lightning hazard in and near a severe thunderstorm. For the test to work, channel 2 must be free of broadcasting stations. The TV test is conducted as follows, in this order:

- Switch the channel selector to the highest empty VHF channel
- Set the brightness control so the screen is almost completely dark
- Move the channel selector to channel 2

Under normal conditions, the screen will stay near-dark, except for momentary bright flashes caused by lightning strokes nearby. If the screen brightens uniformly and more or less constantly, a funnel cloud exists within a few kilometers. This brightening of the screen, almost as if a distant TV station is fading in and out, is the result of the numerous intracloud lightning strokes that occur in tornadoes.

Although it is possible to detect a tornado by listening for its characteristic rumble or by using a radio or TV set as just described, tornadoes can strike even when neither of these classic warning signs is observed. Do not assume that the lack of a loud rumble, or normal behavior of a television set, means there is no threat! The best way to keep informed about tornadoes, when conditions favor them, is to listen to weather announcements on local radio or television stations. The meteorologist has the radar with which hook-shaped echoes can be detected. The police, highway troopers, and other officials will report funnel cloud sightings. Tornado warnings will be disseminated through the commercial broadcast stations. If you hear such a warning, take it seriously! Even a moderate tornado has winds as strong as those in a major hurricane.

MOVEMENT

After a funnel cloud touches down, it is impossible to precisely predict its track on the surface. The average forward speed is roughly the same as that of the storm that contains it, normally 20 to 30 kt (23 to 35 mi/h). As the funnel snakes along, however, it may briefly speed up to 40 kt (46 mi/h) or more, come to a stop for a moment, or even backtrack for a short distance.

Single tornadoes usually describe straight (Fig. 5-2A) or looping (Fig. 5-2B) tracks along the surface. The looping tornado is especially dangerous, because it passes over certain points twice. Some storms generate multiple funnel clouds that move in curved paths and then dissipate. Vortices form near the trailing part of the mesocyclone and move around to the front of the circulation before dying out. This produces a broken track (Fig. 5-2C). The average speed of this type of tornado is a little greater than the speed of the mother cloud. The multiple-funnel tornado is, like the looping type, particularly dangerous, because some places may be hit by two or more different vortices in rapid succession.

The average movement of a tornado is in the same direction as the movement of the mother cloud, and most thunderstorms move generally west-to-east or southwest-to-northeast. But a looping-path or multiple-vortex tornado can strike a specific point on the surface from any direction. If a looping-path tornado (Fig. 5-2B) is contained in a storm moving from the southwest toward the northeast, for example, the funnel will *on average* travel southwest-to-northeast. However, the tornado will hit some places from the south, from the east, or even from the north. Besides this, there are instances in which tornado-containing thunderstorms themselves travel in an unusual directions, such as north-to-south. In hurricanes, tornado-spawning thunderstorms can travel in any direction.

If a tornado or funnel cloud can be seen, its movement is might seem obvious, but it can be deceiving. A straight-path funnel cloud, several miles away and retreating, presents little danger. But in many cases, the forward progress of a tornado is impossible to ascertain because the motion is complex. If you see a funnel cloud, the best thing to do is take shelter until you are certain that the threat has passed.

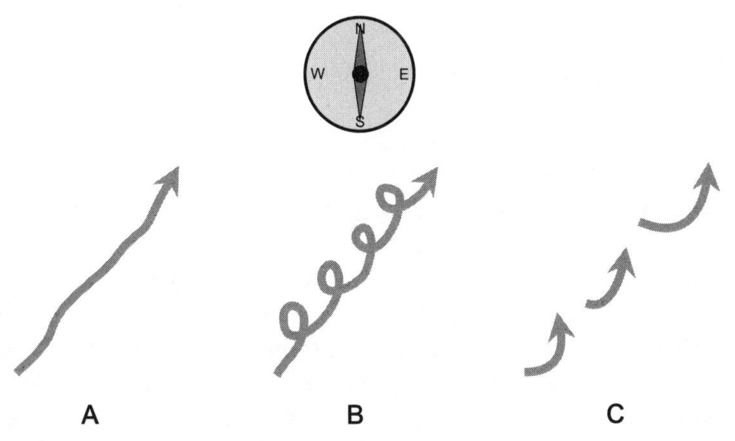

Fig. 5-2. A tornado can take a generally straight (A), looping (B), or broken (C) track along the ground.

PROBLEM 5-1

Is it possible for a straight-path tornado to acquire a looping or multiple-vortex path?

SOLUTION 5-1

Yes. A tornado can change its character or behavior abruptly. This is, in part, why it is a dangerous business to attempt to predict exact points where a tornado will strike.

PROBLEM 5-2

Is it necessary for a large, severe thunderstorm to exist before a tornado can develop?

SOLUTION 5-2

No. Occasionally, tornadoes appear in ordinary showers and thunderstorms. However, such tornadoes are rarely as intense or as large as the ones associated with severe thunderstorms.

Tornado Behavior

Most tornadoes get their work done in a hurry. A typical funnel cloud lasts for less than an hour, and if it becomes a tornado, it stays on the ground for only a few minutes. The average tornado path is only a few kilometers long, but if that path cuts through a heavily populated area, the resulting damage can be great.

FROM BIRTH TO DEMISE

As a funnel cloud first begins to reach toward the earth, it is usually thin. Multiple narrow vortices may appear. Occasionally, however, a wide vortex descends abruptly from a mesocyclone. Many funnel clouds dissipate before they ever touch down to become tornadoes. Of the funnel clouds that do reach the earth, most never attain diameters of more than a few hundred meters.

Once in awhile, a tornado acquires a large diameter, develops winds in excess of 200 kt (230 mi/h), and lasts for an hour or more. Some such "maxi-tornadoes" travel hundreds of kilometers in uninterrupted paths. One tornado that still receives historical attention was the so-called *tri-state twister* of 1925. It killed over 600 people in a rampage lasting for the better part of a day and traveled in a straight-line path at a speed of about 50 kt (58 mi/h).

Once a tornado has reached the surface, the vortex, with its associated updraft, launches all sorts of objects into the air. Dirt, grass, and leaves color the clouds, giving them a bizarre aspect. Plumes of dust may rise up and surround the funnel, making it look bigger than it actually is. As the vortex begins to dissipate, it rises partially back into the mesocyclone, its path of destruction narrowing and the damage becoming less severe. Small tornadoes sometimes vanish abruptly, as if dissolving into the surrounding atmosphere. The base of the mesocyclone swirls more slowly, until finally it is no longer evident. The ragged cloud base, darkened by dirt pulled aloft during the life of the tornado, still presents a threatening spectacle, but the danger has passed.

Once a thunderstorm has stopped producing funnel clouds, it normally weakens and dissipates over an hour or two. Some thunderstorms generate tornadoes for a while, weaken temporarily, and then strengthen again and spawn more funnel clouds. This is most likely to happen in a multiple-cell system or in a squall line.

STRANGE EFFECTS

Tornadoes are known for their awesome, and sometimes unearthly, methods of destruction. Pine needles are driven into wooden planks, timbers hurled through concrete walls, and trailers wrapped around trees. People have become airborne in tornadoes—some inside vehicles or houses, and others all by themselves. Whole houses have been lifted off their foundations and moved several meters without other structural damage. A tornado might reduce a house to a pile of rubble, while the shade trees in the yard don't lose a single branch. A tornado once pulled a curtain through a sealed window around the edge (that is, right through the seal!) without breaking the glass. In another instance, a wooden beam passed through a piece of glass, punching a hole in the glass but not shattering it.

If a tornado passes over unusually colored soil, the funnel and the whole cloud base acquire the tint of that soil. This is responsible for the eerie light that often precedes or accompanies a tornado. Ordinary soil produces a greenish or yellowish hue. Some types of soil or sand can make the sky look pink, brown, red, or even purple. If a tornado passes over a lake, stream, or swamp, the vortex picks up fish and frogs. The creatures are carried by the updrafts within the mother cloud, and later fall back to the earth along with rain or hail. If such a storm passes over a populated area, the streets will be littered with fish and frogs. Tornadoes have been known to strip chickens and other birds of their feathers. Most of the birds are killed as they are flung about by the storm winds, but some survive.

WINDS

In cases where tornadoes directly strike weather stations, the anemometers are invariably wrecked before the speed of the winds can be determined. Indirect methods must be used to ascertain how hard the winds blow in a tornado.

It was once believed that intense tornadoes packed winds of 500 kt (580 mi/h) or more, perhaps exceeding the speed of sound. It was theorized that the characteristic roar of a tornado was caused by air breaking the sound barrier as it whirled around. Analysis of the structural damage caused by tornadoes, however, indicates that the wind speeds are less than was previously thought. It is doubtful that the winds ever exceed 300 kt (about 350 mi/h), but that is fast enough to cause almost complete destruction.

Most people have witnessed winds of minimal hurricane force, and know how much damage can be done by air moving at this speed. A 250-kt wind causes damage that is dozens of times more spectacular and costly than the damage caused by a wind of 60 or 70 kt. In large measure, this is the result of flying debris smashing things to pieces, in turn generating still more flying debris, which in turn tears things up still more—a vicious circle of devastation that can turn a handsome neighborhood into a pile of timbers and cinder blocks within a couple of minutes.

THE FUJITA SCALE

A well-known meteorologist and tornado research scientist, Dr. T. Fujita of the University of Chicago, has formulated a scale of tornado intensity based on observation of the effects of hundreds of the storms. The wind speeds range from less than hurricane force to more than 260 kt (300 mi/h). There are six categories, designated F0 through F5, in order of increasing violence. The *Fujita tornado intensity scale*, as it is called, is outlined in Table 5-1.

In a tornado with winds that rotate counterclockwise, the winds blow fastest in the right-hand part of the vortex, where the forward movement of the tornado adds to the rotational motion of the air. If the winds whirl at 120 kt, for example, and the funnel cloud travels at 20 kt, then the right-hand side will pack 140-kt winds while the left-hand side will contain winds of 100 kt (Fig. 5-3A). This might not sound like much of a difference, but it represents a large factor in terms of destructive power.

The additive wind-speed effect is more pronounced in multiple-vortex tornadoes. The vortices all rotate around a common center. Suppose the center of the tornado moves across the surface at 20 kt, the circulation of the main vortex

Table 5-1. Fujita intensity scale for tornadoes.

Fujita category	Wind speed in knots	Wind speed in miles per hour
F0	Less than 63	Less than 72
F1	63–97	72–112
F2	98–136	113–157
F3	137–179	158–206
F4	180–226	207–260
F5	More than 226	More than 260

occurs at 120 kt, and the *secondary vortices* revolve at speeds of up to 60 kt relative to the circulation of the main vortex. In that case, the maximum wind speed is 120 kt + 20 kt + 60 kt, or 200 kt, as shown in Fig. 5-3B.

Within any tornado funnel, small eddies form. These eddies are known as *suction vortices*, and they can revolve clockwise or counterclockwise. The combined effects of forward motion, main circulation, and secondary-vortex circulation, along with the winds in suction vortices, can be extreme. For example, suction vortices revolving at 50 kt, combined with the winds in the system shown at B, can produce gusts of up to 120 kt + 20 kt + 60 kt + 50 kt, or 250 kt (Fig. 5-3C). Such gusts, in addition to their sheer violence, cause rapid, multiple shifts in the wind direction. This has a wrenching effect on structures that the tornado encounters. The small-scale gusting and wrenching explains the signature bizarre damage that intense tornadoes sometimes leave behind.

PRESSURE

In some ways, a tornado is like a miniature low-pressure system. The barometric pressure inside the funnel cloud is lower than the normal atmospheric pressure. The intensity of a tornado is correlated with the internal pressure.

It is difficult to measure the exact barometric pressure inside a tornado. When a twister bears down, people do not rush to barometers to make observations. The inside of a tornado funnel is far from a perfect vacuum, because people have been at the centers of tornadoes and survived without serious injury. One man who had been through several tornadoes noted that the rapid drop in pressure

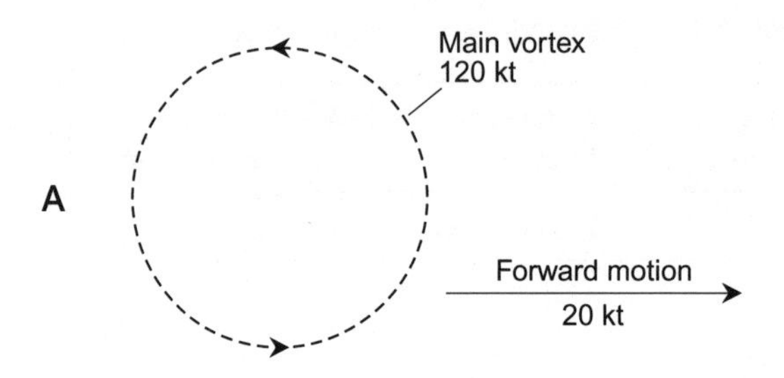

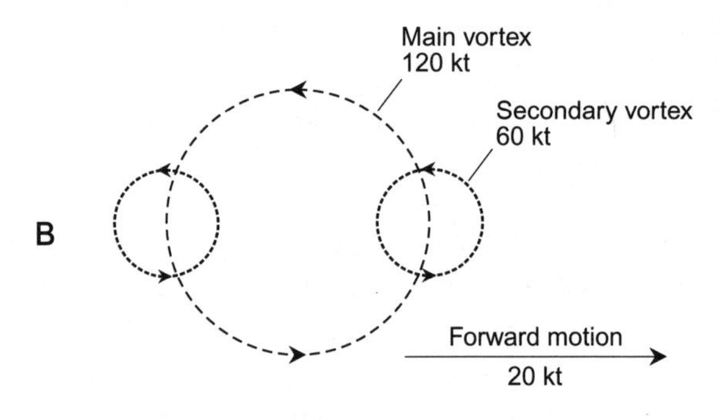

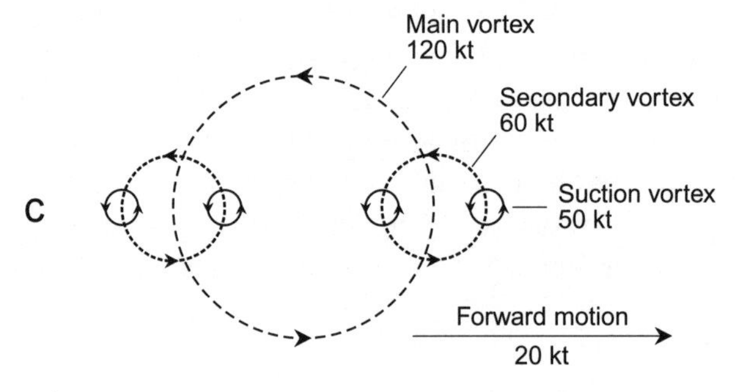

Fig. 5-3. At A, the forward motion of a tornado adds to the wind speed in the main vortex. At B and C, secondary and suction vortices produce extreme gusting and wrenching effects.

caused his ears to pop, as they would when driving up a steep mountain. Some people who have been through tornadoes have noted that they had no difficulty in breathing, even while they were in the center of the funnel.

Barometric pressure does not have to drop very much (in terms of the percentage of normal atmospheric pressure) in order to produce large physical forces. Consider the example of a closed house in the center of a tornado in which the barometric pressure suddenly drops to 90% of normal. This will produce a pressure change sufficient to knock down the walls or lift the roof.

For many years, it was believed that the primary destructive power of a tornado comes from the pressure change as the funnel passes by. It was theorized that buildings explode because of expanding air inside. Today, most meteorologists believe that the winds, and particularly the secondary and suction vortices, are responsible for most of the damage that a tornado inflicts on buildings. Once a building has been broken apart, the debris is scattered by winds in and around the funnel cloud. This debris strikes other buildings and compounds the damage.

The theory that damage is caused by wind, and not the diminished air pressure inside a tornado, is supported by analysis of damage patterns. If the drop in the pressure caused buildings to explode, most of the debris would be carried along with the storm and would fall in the leeward (usually northeast) corner of the building foundation. But this is not what happens. In most situations, the windward side of the foundation is where most of the wreckage lands. This is consistent with what would be expected when a building is simply bulldozed down by the lateral force of the wind.

LOOKING INSIDE

Tornadoes seldom pass over meteorological installations, and this makes it difficult to probe inside the whirling storms. Little is known about conditions within a funnel cloud. There are a few eyewitness accounts by people who have been through a tornado and seen the chamber within the opaque wall of cloud. Those who have had the experience tell of relative silence, a rotating cylinder or "pipe" of clouds, frequent or continuous lightning, and sometimes the smell of ozone gas. Ozone is produced by electrical discharges through the air, such as nearby lightning strokes. To some people, it smells like "natural gas." To other people, it smells like bleach.

In the latter part of the 20th century, the National Oceanic and Atmospheric Administration (NOAA) developed a portable device for measuring extreme wind speeds and atmospheric pressures. It is known as the *totable tornado observatory*, or *TOTO*. The TOTO instrument contains a wind vane for determining wind direction, a pressure-sensitive wind speed detector, an electric charge sensor, a barometer, and a thermometer. The instruments are designed to withstand the winds in all tornadoes. The device is shaped like a barrel, and is weighted to minimize the risk that it will be blown away. Instrument readings are plotted against time, so they can be evaluated when TOTO is recovered.

Putting TOTO in place is a dangerous business. When a tornado is sighted, two scientists drive toward its path in a truck, along with TOTO. An educated guess must be made as to the exact path the tornado will take. The closer the scientists can get to the tornado as it approaches, the better the chance of its passing over TOTO. Once the position has been chosen, the scientists stop the truck, wheel TOTO out and set it down, get back in the truck, and speed away. If the scientists are lucky, TOTO will be struck (and they will not).

PROBLEM 5-3

Suppose a tornado approaches from the west, traveling along a straight-line path. From which direction will the strongest winds blow?

SOLUTION 5-3

The strongest winds will blow from the west—from the same direction as the tornado travels. This is true regardless of the complexity of the internal vortex structure. Multiple westerly wind components will, in certain parts of the system, all add up together.

PROBLEM 5-4

As a tornado approaches, is it possible to predict with certainty which part of the funnel will cross a specific point on the surface, thereby obtaining a forecast of the severity of the storm at that point?

SOLUTION 5-4

In general, this is not possible. Tornadoes can change in intensity, and their paths can change, from moment to moment. In addition, it is difficult to determine the wind speed within the system without the use of Doppler radar.

Variations on a Violent Theme

Does the mention of the word "tornado" bring to mind a dark, funnel-shaped, roaring cloud mass that scours the floor of a prairie, illuminated by eerie light, destroying everything in its path? Some tornadoes fit that description, but tornadoes vary in character. Some occur over water; some are snowy-white or gray in color; some are so thin as to be almost invisible; some resemble an indistinct cloud mass that boils and rolls along.

TROPICAL AND SUBTROPICAL TORNADOES

Large, violent tornadoes do not take place in the tropics and subtropics as often as they occur at temperate latitudes. This is mainly because the battles between warm and cold air masses are less common, and generally less intense, in the tropics and subtropics as compared with the temperate zones, especially over North America. The atmosphere is comparatively free of frontal cyclones at low latitudes, particularly in the summer.

Small and moderate tornadoes are common in the tropics and subtropics. In the extreme southern United States, especially in Florida, there are occasional tornadoes in conjunction with heavy thundershowers. In the winter, cold fronts from the continent move down the state, triggering thundershowers and squall lines, and some of these can spawn tornadoes. In the summer, cumulonimbus clouds build up almost every day as the sun shines practically straight down on the land and heats the moisture-laden air. In the center of the state, the resulting thundershowers generate so much lightning that the region is sometimes called the *lightning belt*. Torrential rains, gale-force winds, and small hail are common in these thundershowers. Small tornadoes are sometimes observed. Australia lies almost entirely in the tropics and subtropics. Tornadoes are observed fairly often there, and they arise from the same causes as tornadoes in Florida. Much of Asia lies in the subtropics, and the wetter parts of that continent sometimes get tornadoes.

Tornadoes have been observed in the outer rain bands of tropical storms and hurricanes. In tropical hurricanes, the tornadoes are usually small, and do less damage than they do in the big Midwestern storms (although this fact is academic in the eyewall of an intense hurricane where the sustained wind blows at tornadic speed anyway). Near a hurricane, and particularly as it moves ashore or encounters cooler air at temperate latitudes, tornadoes present a greater hazard. They can strike virtually without warning and do considerable damage. Funnel clouds are most likely to develop in the forward semicircle of a tropical cyclone, especially the right front quadrant in the northern hemisphere or the left front quadrant in the southern hemisphere. If you are in a hurricane warning region, you should listen to local broadcast stations for tornado warnings.

WATERSPOUTS

Not all tornadoes form or travel over land; sometimes they occur over bodies of water. A twister may pass over lakes and rivers during its trek across the countryside. Then, for a short time, it becomes a *waterspout*. The cloud changes color as the material that is sucked-up becomes liquid instead of solid.

Marine waterspouts form over the warm waters of tropical and subtropical oceans. They are also observed during the warmest months in the temperate zones. Waterspouts are common over the Gulf Stream off the east coast of the United States, especially in the summer. These waterspouts are often white or bluish white, contrasting with the gray cloud bases. There may be two or more separate funnel clouds.

Ocean waterspouts, and the thundershowers that breed them, are rarely as violent as their land-born counterparts, but a small craft can be damaged or capsized by a large waterspout. For this reason, all pleasure-boat owners should be aware of the potential for danger when waterspouts threaten. Large vessels are not usually affected much by waterspouts, but they can present an inconvenience to the captain of an ocean liner or destroyer.

Wind speeds in waterspouts do not often exceed minimal hurricane force. According to one report, a U.S. Navy ship was struck directly by a waterspout and received nothing more than a tropical-storm-force buffeting. Personnel at the front of the ship noted gale winds at starboard (from the right), while personnel aft reported gale winds at port (from the left). Sailors amidships reported sharply shifting gale-force winds. There was no damage to the ship or injury to personnel. Nevertheless, waterspouts should be taken seriously. A person caught unaware can be swept overboard. In addition, loose objects can be picked up and tossed around, creating the potential for injury. Once in awhile, a waterspout can grow to considerable size and acquire wind speeds comparable to F2 or F3 tornadoes.

DUST DEVILS

If you live in an arid region, you have seen little vortices form and dissipate on hot, windy days. The dust begins to stir, and within a few seconds the rotating nature of the "mini storm" is apparent. Leaves, loose papers, and other light objects are picked up and carried along. If the whirlwind strikes you directly, you get dust in your eyes. Wind speeds may reach 50 to 60 kt for a moment. These are *dust devils*.

In the desert, large dust devils can form. This type of vortex is most likely to occur on blistering hot afternoons. The weather is usually clear or partly cloudy. The development of the storm is fueled by rising, heated air. Sand and dust are picked up and carried aloft. This makes it look like a small tornado without a mother cloud. The diameter is a few meters. It is wise to seek cover from an approaching dust devil of this type, because blowing sand and debris can cause personal injury to people who are unprotected and who are directly struck.

Dust devils rarely have winds stronger than gale force, and they almost never do major damage to property. But they can occasionally produce winds strong enough to uproot small trees and shrubs. The maximum life of a dust devil is on the order of a minute or two. Multiple dust devils can appear and disappear spontaneously, each one lasting only a few seconds.

FIRE-GENERATED TORNADOES

Whirlwinds similar to dust devils sometimes result from the intense heat generated by massive fires. During the Allied bombings of German cities in the Second World War, *firestorms* occurred in which the intense heat caused the air to expand and rise. A nuclear burst of several megatons, or the impact of a small asteroid on a land mass, could be expected to generate tornadoes. Forest fires have been known to produce tornado-like whirlwinds with destructive effects. Artificial fires have been used by scientists to generate mini-tornadoes for research purposes.

HILLS AND CITIES

Small tornadoes can be produced by *orographic lifting*. As the wind blows against a hill or mountain, the air is forced upward because it has no place else to go. The rising air can begin to twist, especially if the hill or mountain has irregularities that produce wind shear. The vortex tightens until it resembles a dust devil. A similar effect can occur when air is forced down a slope.

Most of us are familiar with the effects of buildings in a city. It might be a bright, clear day with balmy breezes, and you walk out of a tall building and are buffeted by gusts of gale force. If it is a windy day, the funneling effect of adjacent tall buildings can produce whirlwinds strong enough to blow people off balance. During a thunderstorm or tropical cyclone already packing high winds, the effects of buildings can produce gusts and updrafts characteristic of small tornadoes.

REVERSED SPIN

Although the Coriolis effect suggests that a tornado should rotate counterclockwise in the northern hemisphere and clockwise in the southern hemisphere, there are exceptions. As one man in Kansas watched a tornado pass directly overhead, he noticed that some of the peripheral suction vortices appeared to be spinning clockwise, although the main funnel twisted counterclockwise. Waterspouts have been known to spin in either direction. In some cases, an examination of tornado damage has indicated that the winds must have been spinning clockwise.

Although scientists are not certain why some tornadoes "spin the wrong way," one theory holds that tornadoes sometimes occur in conjunction with downdrafts instead of updrafts. Such whirlwinds would have internal pressure higher than that of the surrounding air, and they would therefore rotate in an anticyclonic sense. The winds would blow outward and downward. Such vortices might develop in the anticyclonic part of a large thunderstorm in connection with a microburst (Fig. 5-4). A tornado of this type would be especially dangerous, because the anticyclonic part of a thunderstorm usually contains heavy rains, which would conceal the funnel cloud (if there was a cloud formation associated with the whirlwind).

THE OMAHA TWISTER OF 1975

As a classic example of a deadly urban tornado, let us consider the Omaha twister of 1975. A cold front moved eastward across Nebraska on May 6 of that year, and it steadily intensified. By the time the clear, cool mass of air behind the front neared Omaha, the line of clouds was sharply defined, and the clouds unleashed a deluge of rain and hail. Funnel clouds formed and touched down in northeastern Nebraska near the South Dakota border. Damage was reported in several small towns. The rain and hail continued to batter Omaha, driven by winds of gale force. At about 4:00 P.M. Central Daylight Time, the storm began to let up, and conditions became strangely quiet. The sky acquired a yellow-green hue. A massive funnel cloud formed and reached the surface.

Civil defense sirens were activated. Amateur and Citizens Band radio operators observed and reported the storm's progress as a second funnel descended, reached the ground with a trundling noise, and proceeded to follow a southwest-to-northeast track into the suburbs of Omaha. People took shelter wherever they could. The storm passed near a hospital and then blew a restaurant to pieces, killing an employee who took refuge in one of the bathrooms. (Although she had chosen one of the safest places in the building, the whole structure was demolished. Only a detached, subterranean storm cellar can provide complete safety from the winds of an intense twister.) The storm moved on, wrecking businesses and homes. Automobiles and trucks flew through the air along with pieces of wood, brick, and turf. The tornado grew until its diameter was in excess of 1 km. The storm skipped over a freeway and turned northward along West 72nd street. Two more people died in the tornado as it roared northward through the western

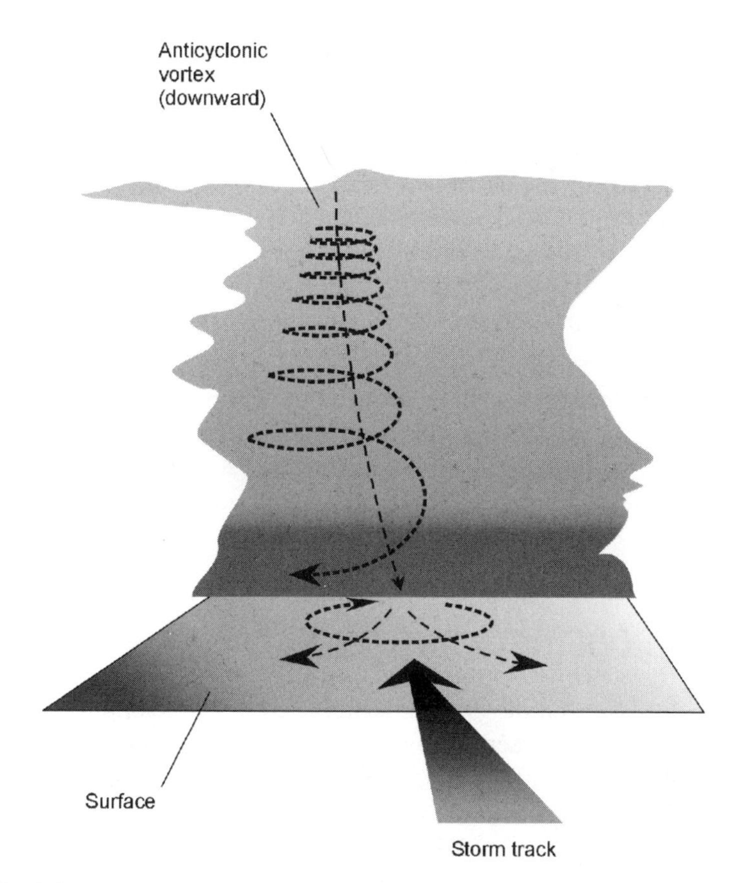

Fig. 5-4. In a severe thunderstorm, an anticyclonic vortex may reach the surface, causing tornadic winds that spin in a direction opposite that of the more common cyclonic tornado.

suburbs of Omaha. One woman was sitting in her living room as her house collapsed. A man was killed outdoors on a street corner. Finally, the funnel narrowed and retreated back into the mesocyclone.

The path of the storm was approximately 14 km (9 mi) long, began on a golf course, ended on another golf course, and destroyed hundreds of buildings in between. The whole ordeal lasted for approximately 15 min. The duration of tornadic winds was less than 1 min at any single point on the surface.

The death toll from the 1975 Omaha twister (estimated at F4 on the Fujita scale) was only three people, in spite of the fact that the storm passed through a densely populated area. Effective tornado warnings saved dozens, or perhaps hundreds, of lives. As the dazed residents of Omaha surveyed the damage of their property, they gave thanks that their lives and health were still intact, but property damage was incalculable. Hundreds of cars and trucks had been wrecked. Metal structures had been bent by the sheer force of the wind. National Guardsmen were called out to keep order and to prevent looting. In the hardest-hit areas, the rubble was piled so deep that people could not walk through it.

Protecting Property and Life

In recent years, meteorologists have determined that the winds of tornadoes do not blow with sonic or supersonic speeds as was once thought. The worst storms generate winds of perhaps 300 kt (330 mi/h). This has given rise to speculation about the possibility that tornado-proof, above-ground structures can be built, at least for storms of F3 force or less.

TORNADO-PROOFING A BUILDING

It is doubtful that large buildings, gymnasiums, arenas, or shopping malls can be constructed with the reinforcement necessary to resist tornadic winds. The massive roofs would be torn off by airfoil effects. Small homes might be designed to withstand F3 or perhaps even F4 winds, but special construction methods and precautions would be necessary.

Experience has shown that the windows are usually the first part of a house to be damaged by high winds. They are either blown out by the force of the wind itself, or else taken out by flying debris. The roof quickly follows. As air flows across a surface, the pressure is reduced. Flat roofs, or roofs with shallow peaks, are the most vulnerable (Fig. 5-5). Steep roofs impede the flow of air and are therefore more resistant to damage. All roofs should be secured with hurricane clamps during the construction process. Attic vents help to equalize the pressure above and below the roof during strong winds.

The orientation of a house affects its susceptibility to damage by winds and debris. A rectangular house will receive the least damage if a short wall faces an oncoming tornado (Fig. 5-6A). In that case, the exposed surface area is small, and the overall force is minimized. The worst situation occurs when the tornado approaches toward a long wall (Fig. 5-6B). Tornadoes in any given place can approach from any direction; but statistically, there is a single direction, usually west or southwest, that represents the mean, or average. Before homes are built in tornado-susceptible regions of the country, construction engineers may wish

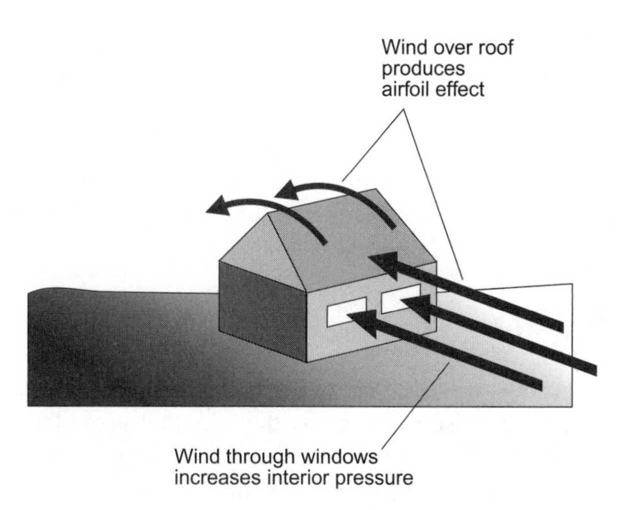

Fig. 5-5. High winds create a drop in pressure above a roof, and can also produce increased pressure within a building. These forces combine to tear the roof loose.

to determine the mean *azimuth* (compass bearing or direction) from which tornadoes approach, and orient new homes accordingly. A single-story house generally has less wall area, and better immunity to wind damage, than a multiple-story house. *Geodesic dome* structures are the most wind-resistant of all.

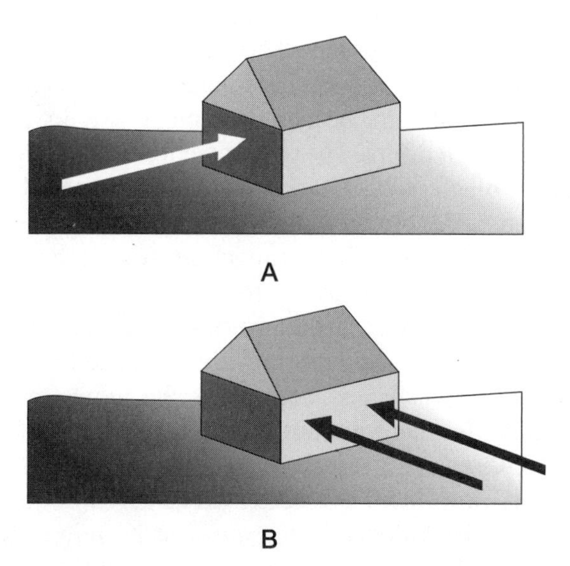

Fig. 5-6. A high wind blowing against a short wall (A) exerts less destructive lateral force on a house than the same wind blowing against a long wall (B).

Most experts believe that it is a waste of time to attempt to build a conventional home that will endure a direct hit by an F5 tornado. Such severe twisters rarely strike any single place, even in the heart of "tornado alley." The chances of any given house getting hit by an F5 tornado are vanishingly small. For absolute full protection against a direct hit by the most intense possible tornado, the entire house—roof, garage, and all—must be located underground.

PERSONAL SAFETY

Tornadoes are dangerous, but they cause fewer injuries and deaths, on the average, than lightning. One reason for this is the fact that tornadoes batter only a tiny percentage of the land area of the world, while lightning can occur almost anywhere. In addition, people are likely to run for shelter as a tornado bears down, but are rarely fearful when there is "nothing but a little bit of thunder." Tornadoes command more respect than lightning because of their spectacular appearance, the terrible sounds they can make, and the fact that they can inflict unbelievable damage.

In the United States, the largest number of tornadoes strike in the Midwest and South. The Great Plains states are particularly hard hit, but interestingly, the greatest number of people are killed in the South, especially in Louisiana, Mississippi, and Georgia. There are two factors that are believed to be responsible for the greater number of tornado-related deaths in the South as compared with the Midwest, in spite of the fact that the Midwest gets more tornadoes. First, in the South, many houses lack basements, either because of low-lying ground and a high water table, or because basements are not traditional in that region. Because a basement offers superior protection in a tornado, a house without a basement poses much more danger than a house with a basement. Flying debris tends to be propelled horizontally in straight paths. For this reason, a depression, such as a basement, offers protection. Also, a house with no basement can be swept completely away by the winds of a tornado. The second reason that tornado deaths occur disproportionately in the South is a comparative lack of warning systems and storm awareness. The tornado warning system in the Midwest is sophisticated because of the more frequent F4 and F5 storms that occur there. In the rural South, warning systems are less prevalent.

The principal life-threatening aspect of a tornado is the flying and falling debris. The temporary drop in air pressure is not likely to injure a person. The chances of getting hurt or killed in a tornado can be minimized by avoiding places where objects are likely to fall or be hurled through the air.

TAKING COVER

The safest place to be, should a tornado directly strike in your vicinity, is several meters underground in an old-fashioned *storm cellar*. This type of shelter is completely separate from the main living structure and has an entrance like a trap door, made from heavy steel or reinforced hardwood. Another excellent refuge is a *fallout shelter*. These heavy structures are practically immune to demolition from anything less than a direct hit by a nuclear bomb. Because the storm cellar is completely autonomous, people inside are totally safe from flying debris. Unfortunately, most homes built today, even if they have basements, have neither storm cellars nor fallout shelters.

From studying the damage done by tornadoes, scientists have found that the windward side of a structure is the least safe, and the leeward side is the most safe. This is contrary to the earlier, and still widely believed, idea that the southwest comer of the basement is the best place to take cover. The windward side is considered to be the wall or corner that faces in the direction from which the tornado is approaching, usually the west or southwest. The leeward sides of a structure usually face toward the east or northeast. These windward and leeward directions are, however, by no means absolute. A June 1971 storm that struck near Rochester, Minnesota, moved from north to south. In that storm, the southern parts of basements would have been the safest had the funnel clouds touched down in the city. It is important that you know the general direction from which a tornado is approaching.

The windward part of a building is the most dangerous, because the greatest wind speeds are found in the part of the tornado in which the forward movement adds to the whirling effect. In a tornado approaching from the southwest, therefore, the highest winds blow from the southwest. Thus, most of the debris approaches from that direction, and is hurled at the highest speeds from the southwest toward the northeast. The windward walls of a building, if not pushed down outright, are the most susceptible to penetration by flying objects. Even in the basement of a house, the windward corner can be dangerous. The whole house may be picked up and moved a few meters, with the windward side or corner dropping into the basement.

If you can get to a basement, you are better off than you are if you must stay on a floor that is above ground level. If there is no basement, you should go to the first floor on the leeward side or in the center of the building, and crawl under something heavy such as a bed, desk, or table. If you are in a large building, you should go to the basement if possible, and get inside a closed room. If there is no basement, you should go to a room with no windows, such as a closet, storage area, or lavatory. Never take shelter in a subway or hallway. They can become wind tunnels, and projectiles can be propelled through them with deadly speed.

PROBLEM 5-5

Suppose you are caught outdoors and a tornado approaches. You can't get to shelter. What should you do?

SOLUTION 5-5

Lie face down in the nearest ditch, ravine, or depression (no matter how shallow), and wait until you are absolutely certain that the threat has passed. "Hug the earth" with your feet facing in the direction from which the storm is coming. Because of friction between moving air and the surface, high winds are usually much less intense within approximately 0.3 m (1 ft) of the surface than at higher levels.

PROBLEM 5-6

If you live in a mobile-home community in a tornado-prone part of the country and there is no community storm cellar available, what should you do?

SOLUTION 5-6

Make every possible effort to motivate the members of the community to underwrite the cost of building a subterranean shelter immediately! In addition, contact your local civil defense authorities to help you implement a plan of action to be carried out whenever a tornado warning is issued.

Quiz

This is an "open book" quiz. You may refer to the text in this chapter. A good score is 8 correct. Answers are in the back of the book.

1. A looping-path tornado is especially dangerous because
   (a) it contains higher winds than other types of tornadoes.
   (b) it is larger in diameter than other types of tornadoes.
   (c) it contains multiple mesocyclones.
   (d) it can strike a specific point more than once.

2. The winds in tornadoes can blow at speeds of up to about
   (a) 75 kt.
   (b) 150 kt.
   (c) 300 kt.
   (d) 600 kt.

3. Suction vortices in a tornado system
   (a) produce strong straight-line winds.
   (b) are accompanied by powerful downdrafts.
   (c) always rotate in an anticyclonic sense.
   (d) can cause strange and extreme damage.

4. Suppose you are building a new home. The home is to be rectangular in shape, with floors measuring 10 m by 20 m. After consulting the historical records, you determine that the mean azimuth (compass bearing) from which tornadoes have approached in the past 100 years is due west. In order to make the home as tornado-resistant as possible, the 20-m-wide outside walls should face toward the
   (a) east and west.
   (b) north and south.
   (c) northwest and southeast.
   (d) northeast and southwest.

5. If a tornado approaches from the south, the most dangerous (that is, the least safe) side of the building in which to take cover is the
   (a) south side.
   (b) east side.
   (c) north side.
   (d) west side.

6. If the main vortex of a tornado rotates at 120 kt, there are secondary vortices rotating at 40 kt, there are suction vortices rotating at 50 kt, and the whole system is moving at 20 kt with respect to the surface from the southwest towards the northeast, the highest winds blow from
   (a) the southwest.
   (b) the northwest.
   (c) the northeast.
   (d) all directions equally.

7. If the main vortex of a tornado rotates at 120 kt, there are secondary vortices rotating at 40 kt, there are suction vortices rotating at 50 kt, and the whole system is moving at 20 kt, with respect to the surface from the southwest towards the northeast, the highest winds blow at
   (a) 120 kt.
   (b) 160 kt.
   (c) 210 kt.
   (d) 230 kt.

8. An approaching tornado can often, but not always, be detected by
   (a) the image of a mesocyclone on a radar set.
   (b) the brightening of a television screen tuned to a vacant channel.
   (c) a noise that sounds like a passing freight train.
   (d) Any of the above

9. On a spring or summer day in an American Midwestern location, you should be aware of a greater-than-normal chance for tornadoes to occur if
   (a) the barometer is falling and the wind is blowing from the south or southeast.
   (b) severe thunderstorms appear to be developing.
   (c) a severe thunderstorm watch has been issued.
   (d) Any of the above

10. A tornado warning will be issued when
   (a) conditions are favorable for tornado development.
   (b) a funnel cloud has been sighted.
   (c) a severe thunderstorm is approaching.
   (d) the barometer is falling.

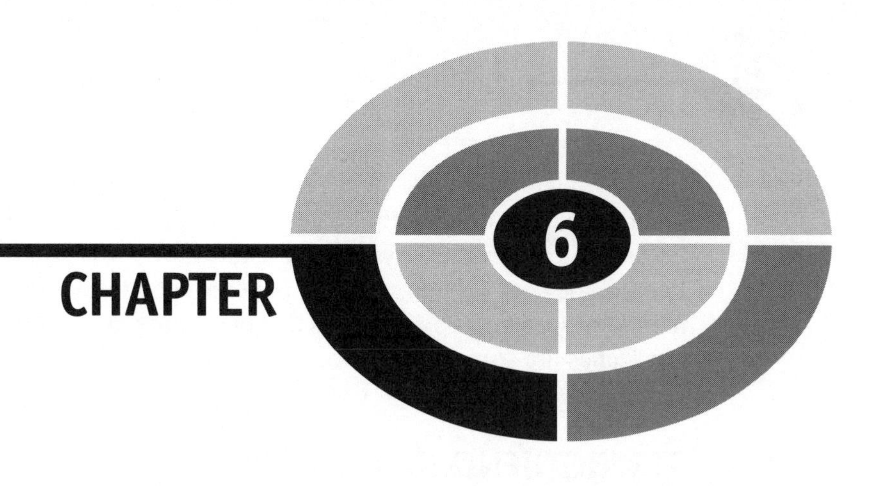

Tropical Cyclones

Hurricanes are the most widely publicized of all storms. They occur every year in both the northern and southern hemispheres. These storms form over the warm waters of oceans in the tropics and subtropics, but they often end up in the temperate regions. In this chapter, we will see how tropical cyclones form, where they occur, what they do, and how people can minimize the dangers posed by tropical cyclones.

A Hypothetical Storm

The date: September 10. The time: 6:20 P.M. Eastern Standard. The weather forecast contains a statement that a *tropical depression* is under scrutiny. It is located in the far eastern Atlantic and shows signs of strengthening. By 11:00 P.M. that day, the depression has become a *tropical storm* with maximum sustained winds (based on a 1-min average) of 34 kt (39 mi/h). It is the fourth tropical storm of the year in the North Atlantic, and is given the name Debby. The following morning, Debby is rapidly gaining strength. The center of circulation is specified as 11.5°N and 40.4°W, far out in the open sea. Debby is, at the moment, a threat only to shipping interests.

In the North Atlantic, the eastern North Pacific, and the central North Pacific oceans, intense tropical cyclones are known as *hurricanes.* In the western North Pacific, they are called *typhoons.* Severe typhoons are called *supertyphoons.* In the Indian Ocean, tropical storms and hurricanes are called *cyclones.* In Australia, hurricanes are sometimes called *gales.*

STRENGTHENING

By the evening of September 11, Debby packs maximum sustained winds of 64 kt (74 mi/h), and has thus become a *minimal hurricane.* Conditions favor further strengthening. The storm is moving almost directly westward at 18 kt (21 mi/h). The television newscasters show a satellite photograph of Debby. The storm appears as a pinwheel of clouds with a small black dot near the center: a defined eye. We will now keep ourselves informed of the progress of Debby, as will millions of residents of the eastern and Gulf coasts of the United States.

Atlantic hurricanes pose a perennial threat to the United States, as well as Mexico and much of Central America. The eye of Debby, at its present low latitude, moves in an almost westward direction, but it is almost certain that the storm will eventually *recurve* toward the northwest. By the morning of September 12, Debby has the full attention of the National Hurricane Center. She has strengthened with phenomenal rapidity, reminiscent of Hurricane Allen in 1980. The maximum sustained winds are estimated at 150 kt (about 175 mi/h) with gusts to 175 kt (about 200 mi/h). On a scale of intensity from 1 to 5, Debby is a strong 5.

By the evening of September 12, the eye of Debby is located near 12.0°N and 52.9°W. The islands of Grenada, St. Vincent, St. Lucia, and Barbados await the arrival of Debby's fury as seas reach 20 m (66 ft) and breakers pound with unrelenting violence on the windward beaches. The normally gentle trade winds shift to the north, and acquire an ominous gustiness. Then, one after the other, the islands are cut off from civilization.

HORROR STORIES

The first reports from Barbados and nearby islands trickle out from emergency-powered amateur ("ham") radio stations, a mode of communication that always continues to function when conventional modes fail. Almost every building on these islands has been flattened or gutted. There are reports of human casualties.

The hurricane track begins to recurve. The *eyewall,* or most destructive part of the hurricane, is about 160 km (100 mi) in diameter. St. Lucia gets a brief rush of high winds and heavy rain, but escapes the brunt of the storm. The hurricane has slowed its forward motion to 13 kt (15 mi/h) as she lumbers into the Caribbean, heading in the general direction of Haiti and the Dominican Republic. Further reports come from Barbados, which received a direct hit. Trees have been stripped of their bark and leaves; the landscape is barren and desolate; it looks as if there has been a nuclear war. Homeless people wander through the debris.

The fate of the Gulf coast and eastern seaboard of the United States now depends on a number of factors. Atmospheric *steering currents,* which direct hurricane paths, are difficult to predict. Computers at the National Hurricane Center in Miami, Florida, work around the clock, continually updating the status and position of Debby. Sustained winds are near 150 kt (about 175 mi/h). The eye is about 24 km (15 mi) across. Hurricane-force winds cover a roughly circular area with a diameter of 240 km (150 mi).

The hurricane slams into the island of Hispaniola with sustained winds so violent that all the measuring apparatus is destroyed. Residents of Florida prepare for a potential onslaught. Everyone on the Gulf coast also watches the progress of the storm. The high hills of Hispaniola weaken Debby. As she emerges into the northern Caribbean, her intensity has diminished. Weather forecasters know this is typical.

FLUID FORCE

The strongest winds in a hurricane are difficult to measure because instruments rarely survive. The device generally used to measure wind speed is the *anemometer* (Fig. 6-1). This device spins as the wind is caught in the concave parts of the cups. The greater the wind speed, the faster the device rotates. The rotating shaft is connected to an electric generator or tachometer that measures the angular speed and translates this into knots. If the wind is too strong, the assembly can be damaged or destroyed. The highest sustained winds in the most intense hurricanes are thought to be around 175 kt (200 mi/h), with gusts approximately 25 to 30 kt (30 to 35 mi/h) higher than that.

The force produced by the wind in an intense hurricane such as Allen (1980), Gilbert (1988), or Andrew (1992) is difficult to imagine for people who have not been through the experience. In a typical severe thunderstorm in the midwestern or eastern United States, sustained winds of minimal hurricane force (64 kt or

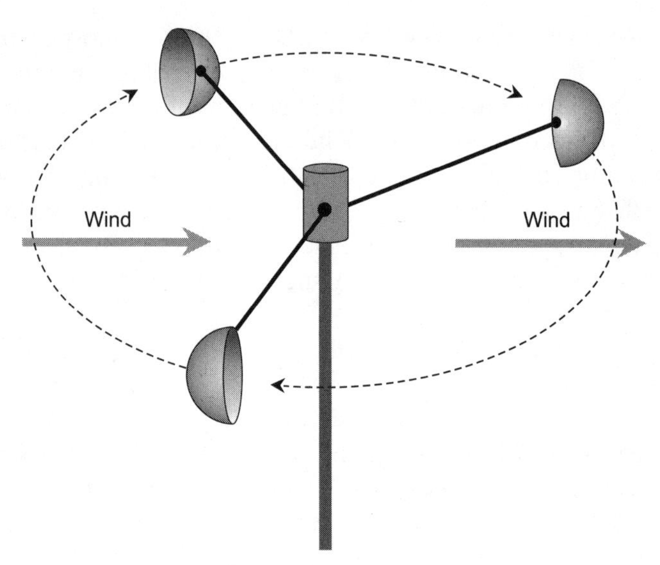

Fig. 6-1. An anemometer cup assembly rotates as the wind blows past. The rate of rotation (angular speed) of the device is proportional to the wind speed.

74 mi/h) sometimes occur. In such storms, trees are uprooted, windows are blown out, some roofs are damaged, utility wires are blown down, small aircraft are overturned, and a few automobiles are wrecked. Wind speeds in the most intense hurricanes are a little more than twice this, but the resulting damage is many times greater. The damaging effects of high winds increase much more rapidly than the numerical wind speed.

The most intense part of a hurricane occurs over a rather small area, usually less than 300 km (185 mi) in diameter. The forward progress of a typical hurricane is only about 10 to 15 kt (12 to 17 mi/h). Therefore, even a fast-moving hurricane will subject some places to high winds for several hours. A large hurricane, if it is slow-moving, can pound a single surface location with high winds for more than 24 h.

WHERE WILL DEBBY GO?

Common sense suggests that there ought to be some way to foretell, at least approximately, where our hypothetical storm called Debby will go, whether it will make landfall on the continent, and if so, where. The art of hurricane-path forecasting is improving continually, but there is still plenty of uncertainty.

Hurricanes that originate in the far eastern Atlantic can eventually strike the coast of the Western world anywhere from Honduras to Newfoundland; many miss North America altogether and dissipate in the chilly waters of the far North Atlantic.

The path that a hurricane takes depends on many factors. One significant influence on tropical storm tracks can be inferred from the climate-control function they serve. The tropics receive more heat from the sun than they radiate into space. The polar regions radiate more heat than they get from the sun. The difference must be made up somehow, in order to prevent a runaway thermal imbalance in which the equatorial regions of our planet would become like Venus and the poles would become like Mars! Hurricanes are a major vehicle for the transfer of thermal energy from the tropics toward the poles. This is why many hurricanes tend to recurve northward in the northern hemisphere, and southward in the southern hemisphere.

Meteorologists are constantly working to improve their methods of tracking hurricanes and predicting their paths. The same basic principles apply to all hurricane regions in the world. The arrangement of continental weather systems, the presence of land masses, and the temperature of the ocean each play a crucial role in the ultimate track that a hurricane will follow. These variables also determine where a storm will form, how it will evolve, how long it will last, and how it will ultimately dissipate.

PROBLEM 6-1

Can North Atlantic hurricanes ever miss the Americas, recurve into the far North Atlantic, and then strike Europe or England?

SOLUTION 6-1

This is rare, but it has happened. Autumn gales that occur in the rural western part of England can sometimes be traced back to tropical origins.

The Azores-Bermuda High

At all times of the year, but especially during the summer in the northern hemisphere, the prevailing winds over the North Atlantic revolve clockwise around a persistent region of high pressure. Because this system is usually centered near the Azores or the island of Bermuda, it is often called the *Azores-Bermuda high,* or simply the *Bermuda high.* This system is responsible for much of the weather that occurs along the east coast of the United States in the summer. In other parts of the world, similar tropical or subtropical high-pressure systems exist over the oceans. In the northern hemisphere, the circulation around these systems is in a clockwise direction; in the southern hemisphere it is counterclockwise.

CONVERGENCE OF THE EASTERLIES

Near the equator, the belts of tropical easterly winds meet. This is the intertropical convergence zone (ITCZ), which was introduced in Chapter 2. It is a region of low pressure and light winds. Thundershowers often form in this zone. In ancient times, this region was called the *doldrums,* because sailing ships passing through it on their way from the northern hemisphere to the southern, or from the southern hemisphere to the northern, were often becalmed for lack of wind. In the Atlantic, the ITCZ almost always lies north of the equator. In August and September, the ITCZ reaches its northernmost position.

Hurricanes can develop when the ITCZ gets far enough from the equator to allow significant Coriolis rotation. In the northern hemisphere, this rotational force takes place in a counterclockwise direction. In the southern hemisphere, it occurs in a clockwise direction. Because the ITCZ is almost always north of the equator in the Atlantic, hurricanes are rarely observed in the South Atlantic. (In March of 2004, the first South Atlantic tropical cyclone on record occurred. It struck Brazil and caused considerable damage. Some scientists suggest that global warming may result in more frequent South Atlantic cyclones in the future.)

WAVES, DEPRESSIONS, AND STORMS

All hurricanes begin as *tropical waves.* These are first observed as irregularities, or "bumps," in the tropical isobars (Fig. 6-2). Showers and thunderstorms develop on the eastern, or trailing, side of the disturbance. Tropical waves, also called *easterly waves,* are common during the summer months, and they usually move from east to west without intensifying. But when conditions are favorable for intensification, an easterly wave can develop a cyclonic circulation. When some of the isobars become closed curves, indicating that a cyclonic circulation has developed, the wave becomes a tropical depression.

A tropical depression is an area of low pressure, similar to a temperate low but without cold fronts or warm fronts. Such a disturbance can continue to strengthen. When this occurs, the central pressure keeps falling, and the wind speed increases. Warm air near the center tends to rise, and the surface winds

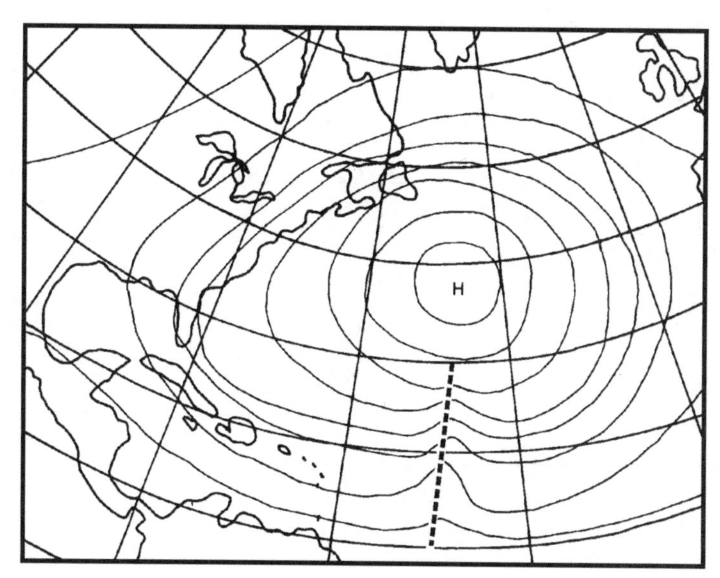

Fig. 6-2. A tropical wave appears as a "bump" in the isobars on the equatorial side of an oceanic high-pressure system.

spiral inward. When the maximum sustained wind reaches 34 kt (39 mi/h), the disturbance becomes a tropical storm, and meteorologists give it a name. If intensification continues until the sustained wind reaches 64 kt (74 mi/h), the storm officially becomes a hurricane.

HURRICANE "BREEDING ZONES"

Hurricanes form just north or south of the equator in many parts of the world (Fig. 6-3). At the equator, there is no Coriolis force, and the big whirlwinds cannot develop; the latitude must normally be greater than 5° either north or south. Hurricanes almost always mature in the tropics, but a few mature in the temperate zones. The key is ocean temperature, which must be above approximately 27°C (80°F) in order for a storm to reach hurricane strength. Hurricanes that threaten the eastern United States mature in the Atlantic Ocean, the Caribbean Sea, or the Gulf of Mexico.

In the northern hemisphere, the hurricane season begins on June 1 and ends on November 30. In the southern hemisphere, the season is reversed. In some parts of the world, hurricane seasons occur at other times; a good example is the Bay of Bengal, where storms occur frequently in April and again in October. The

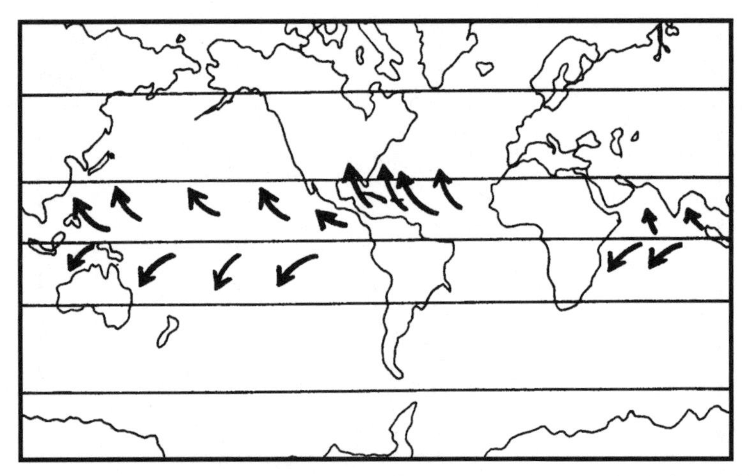

Fig. 6-3. Hurricane "breeding zones" and common tracks throughout the world.

greatest frequency of hurricanes in most of the northern hemisphere is in the months of August and September. This is also the time when a severe storm, such as our hypothetical Hurricane Debby, is most likely to develop.

At the beginning of the Atlantic hurricane season, storms usually mature in the southwestern part of the Caribbean Sea and in the Gulf of Mexico. By late July, the temperature of tropical Atlantic waters has warmed up sufficiently to allow storms to intensify in that region. By mid-October, the main area of development returns to the southwestern Caribbean and the Gulf of Mexico.

In some years, several hurricanes form in the Atlantic, the Caribbean, and the Gulf of Mexico. In other years, there may be few or no storms that reach hurricane intensity in that region. Meteorologists have found some correlation between unusual warming of the equatorial Pacific off the coast of South America (the *El Niño* phenomenon) and a lack of hurricane development in the Atlantic and Caribbean. However, El Niño events seem to cause an increase in the frequency and severity of hurricanes in the western Pacific.

Once a hurricane has formed, it has a natural tendency to move away from the equator, although this does not always happen. The precise path of a particular storm depends on weather conditions over the continents and the locations of the high and low pressure systems at the higher latitudes. Sometimes severe storms actually turn toward the equator before dissipating.

PROBLEM 6-2

Give an example of a major hurricane that turned toward the equator before doing most of its damage.

SOLUTION 6-2

Hurricane Mitch of 1998 intensified over the Caribbean and then turned south, spending most of his energy over Honduras. After weakening, he moved north and northeast causing some wind and rain in South Florida, and finally dissipated over the North Atlantic.

Anatomy of a Hurricane

Although hurricanes are low-pressure systems, they differ from the lows of the temperate latitudes. The hurricane is more symmetrical than the cyclonic storms familiar to inhabitants of Europe, the interior United States, and other temperate regions. The hurricane, in its tropical stage, does not contain frontal systems. The isobars are almost perfect circles, especially in a well-developed storm (Fig. 6-4). The central pressure is usually lower (and sometimes much lower) than that of a cyclone in the temperate regions. When a hurricane contains nearly circular isobars near its center, it is said to have *tropical characteristics.* As

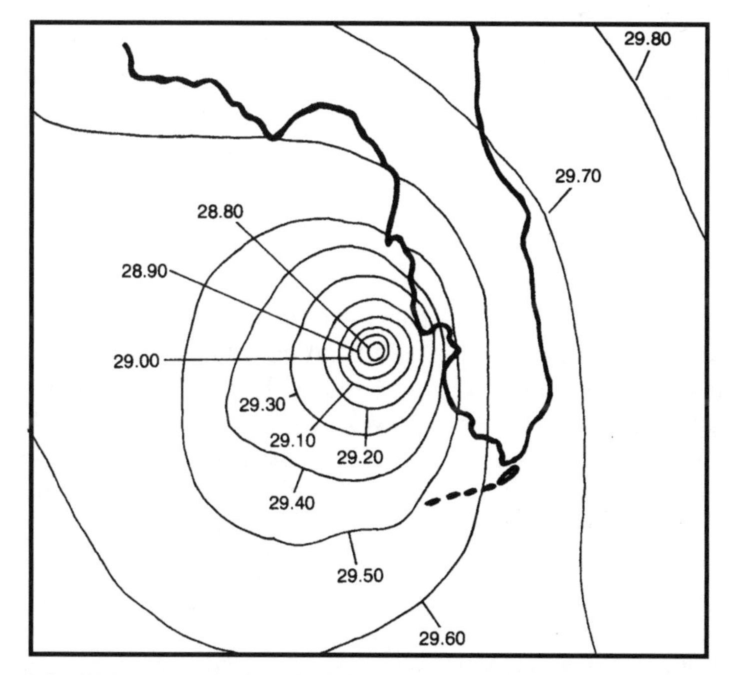

Fig. 6-4. Isobars around a hypothetical hurricane near the Gulf Coast of Florida. Barometric pressures are in inches of mercury (inHg), a unit commonly used in weather broadcasts to the general public in the United States.

a storm moves over land or over cold water, fronts develop and the isobars are no longer circular near the center. Such a hurricane is then said to have become *extratropical.*

CLOUDS, WINDS, AND RAIN

At the periphery of a hurricane, cloud circulation becomes noticeable. This is evident in satellite imagery. The winds in this region are moderate. Rain showers and thunderstorms occur, but they are rarely severe. Three cloud decks can be identified in the outer circulation of a hurricane. The lowest layer, consisting of nimbostratus clouds, produces rain. The middle layer consists of altocumulus and altostratus clouds, and the upper cloud layer is composed of cirrus and cirrocumulus.

Nearer the center, thick clouds form in spiral-like formations. Heavy rain is produced in these regions, which are called *rain bands.* The rain brings high winds to the surface, often attaining gale force. Within 50 to 150 km of the center, the wind circulation picks up rapidly. The *pressure gradient,* or rate of change of barometric pressure per unit radial distance, is steepest in the region immediately surrounding the eye of the storm. This region is called the *eyewall.*

At the surface, the eyewall of the storm produces torrential, almost continuous, rains and violent winds. By the time Debby has reached its peak of intensity, the sustained winds are about 175 kt (200 mi/h). Hurricanes on earth almost never get any more violent than this.

THE EYE

In the central core of the hurricane, the winds and rains abate. This region, which has the general shape of the hole in a bagel, is called the *eye* of the storm. It can be as small as about 8 km (5 mi) or as large as about 160 km (100 mi) in diameter, although most hurricanes have eyes that are 16 to 48 km (10 to 30 mi) across. In the eye, the cloud cover is light. The sky may be partly clear. During the day, muted sunlight filters through, and at night, the stars or moon are sometimes seen. The barometric pressure is lowest within the eye. The level of this pressure is one basis for determining the intensity of the hurricane. While normal air pressure is about 1000 mb, the pressure in a hurricane often drops below 950 mb, and can sometimes get below 900 mb.

The eye of a well-formed hurricane is spectacular, even in satellite images. Pilots who fly "hurricane hunter" aircraft, and the scientists who travel with them, report that the cloud formations give the feeling of being inside a huge, round stadium.

The surface winds of the hurricane spiral inward toward the eye. As the air gets close to the edge of the eye, the angle of inflow gets small, so the winds blow in an almost perfect circle. As Debby approaches from the southeast, the first strong winds will come from the east-northeast (ENE) and then veer toward the northeast (NE).

If you stand at any point and face the eye, the wind blows from left to right in the northern hemisphere and from right to left in the southern hemisphere. This principle of storm circulation is well known to mariners, but it often fools people inexperienced with the storms. Some people think that a hurricane travels in the same direction as the wind blows. If the eye passes over, they think the storm has "blown itself out." Then, when the winds return from nearly the opposite direction, such people will say that the storm "came back."

PROBLEM 6-3

Can destructive winds in a hurricane occur in the direction opposite to the movement of the storm?

SOLUTION 6-3

Yes. In certain locations during intense hurricanes, damaging winds can come from a direction contrary to the forward progress of the system. This occurred in 1992 in extreme South Florida during Andrew. The movement of the storm was from east to west, but the extreme intensity produced sustained westerly winds of 100 kt (115 mi/h) in some locations.

Hurricane Life Cycles

Hurricanes in the North Atlantic and Caribbean usually begin on an east–to–west or southeast–to–northwest track, following the general direction of the prevailing winds. As a storm moves along, it may recurve toward the pole. A hurricane will almost always recurve if it enters the westerlies in the temperate latitudes. Recurving storms threaten primarily the eastern coast and the Gulf Coast of the United States. Non-recurving Atlantic, Caribbean, and Gulf storms affect mainly the shores of Texas, Mexico, and Central America. Some hurricanes recurve harmlessly into the far northern Atlantic. There, the storms combine with temperate low-pressure systems. The forward speed increases, and can reach 40 kt or more. Some hurricanes maintain strong winds well into the temperate zone, particularly in the righthand semicircle where the rapid forward motion of the storm adds to the speed of the revolving winds.

EFFECTS OF LAND AND COLD WATER

A hurricane must eventually face either demise by landfall or demise by cold water. When a storm strikes land, it diminishes in intensity because the winds encounter more friction at the surface than is the case at sea. This effect is most pronounced over hilly or mountainous terrain. When a hurricane moves over cold temperate waters, its supply of energy is cut off, and the storm loses intensity.

After a tropical hurricane crosses a large island and gives up some of its violence, it can be expected to intensify again when it gets back over water. Small islands have little or no effect on hurricane intensity.

FACTORS THAT AFFECT HURRICANE TRACKS

Upon analysis of past hurricane paths and their relationship to the surrounding weather systems, a correlation can be found between hurricane tracks and the conditions in the temperate zone.

When a large area of low pressure exists to the west of a hurricane, the storm will at first be drawn toward this low-pressure region. Then the hurricane will be steered around the eastern edge of the low, where the winds come from the south. The hurricane moves northward into the belt of prevailing westerlies and finally turns to the northeast. The low-pressure area that causes this recurvature may be a broad tropical wave, a cyclone originating in the temperate zone, or another tropical cyclone.

Another scenario that often results in hurricane recurvature is the presence of a low between two highs on or near the continent. A hurricane in the Gulf of Mexico, for example, might turn northward and follow a break between two highs. This brings the storm into the belt of prevailing westerlies. It turns northeastward as it weakens over the land mass.

Still another possible recurvature situation exists when a break occurs in the Bermuda high. It is difficult to predict when such a break will occur, and if it does happen, it might not be well-defined enough to affect the track of a hurricane. But in some instances, storms follow such troughs. These conditions are often responsible for steering hurricanes near Bermuda and the Azores. Under these conditions, a hurricane can maintain much of its strength far into the northerly latitudes.

When the Bermuda high is especially large or strong, or is located somewhat to the west of its usual position, hurricanes normally do not recurve. The zone of the easterlies, or trade winds, expands farther north than usual. When this happens, hurricanes tend to travel in almost straight westerly or west–northwesterly paths. These nonrecurving storms can strike Central America or the Yucatan Peninsula. Occasionally they move into the Gulf and threaten Texas or Mexico.

The prevailing westerlies, which constantly fan the North American continent, sometimes slacken as a result of a massive high over the Great Plains or the southern United States. This kind of situation can prevent hurricanes from recurving. In 1980, Hurricane Allen was a nonrecurving storm. He eventually hit the Texas coast in a relatively unpopulated area. Allen was kept from recurving by high pressure to the north.

Hurricanes are steered, to a large extent, by the winds in which they are embedded. Forecasters can get a good idea of the future path of a tropical storm by locating the isobars over the North Atlantic and the North American continent. A hurricane usually, but not always, moves parallel, or nearly parallel, to the isobars in its vicinity. A fast-moving storm follows a more predictable path than a stalled or sluggish one.

QUIRKY HURRICANES

Tropical cyclones sometimes defy all attempts at path prediction. They occasionally travel against the surrounding winds or across the isobars. A storm may stop, do a "loop-the-loop," and turn in an unexpected direction. In 1935, a hurricane in the Atlantic, apparently bearing down on New England, suddenly veered toward the southwest and struck Miami, Florida. Residents called this storm, having come from the northeast, the "Yankee hurricane," a name that lives to this day. Another example of a hard-to-predict storm was Betsy of 1965. She appeared to be heading for New England or the open sea, and then looped around before ravaging southern Florida, the Keys, and finally the central Gulf Coast.

Another odd quirk in hurricane paths is the occasional tendency for an intense storm, such as Allen, to behave almost as though it has a loathing for land. As Allen churned through the Atlantic and Caribbean, his central vortex never made a direct hit on any of the major islands. Jamaica got the closest shave, but Allen's eyewall veered slightly away from the island, and Jamaica was spared the worst of his fury. As Allen bore down on Texas, moving at a steady clip of about 17 kt (20 mi/h), catastrophe appeared inevitable. The winds along the coast rose to gale force, trees began to blow down, and electric utilities failed. The residents expected an 8-m (25-ft) storm surge and deadly winds. Then Allen stopped short of the coast and spent himself over the waters of the

Gulf of Mexico. When Allen finally did move inland, he had weakened, and did little damage.

THE ROLE OF THE COMPUTER

In recent years, hurricane forecasters have increasingly employed computers to improve the accuracy of storm-path predictions. In the time since records were first kept, many hurricanes have moved into the United States from the Atlantic and the Gulf of Mexico. Their paths have been accurately recorded. A computer can be programmed with all available past data. Then, when a hurricane approaches, the coordinates are constantly fed into the computer. In most instances, several previous hurricanes followed paths (up to the point of current observation) nearly identical to that of the storm at hand. The computer "knows" where these past storms went, and it "knows" all of the surrounding weather conditions and how they are likely to affect the path of the hurricane this time.

When a storm threatens, the current conditions, along with past statistics, are processed by the computer. It gives a mean (average) expected path prediction for the next 12, 24, 48, and 72 hours. The computer also indicates the likelihood that the storm will deviate, either to the left or to the right, from the mean expected path by various amounts prior to landfall. Watches and warnings are issued, according to this data, for a specific section of the coastline.

PROBLEM 6-4

What is the difference between a *hurricane watch* and a *hurricane warning* for a specific location or set of locations?

SOLUTION 6-4

A hurricane watch means that hurricane conditions (sustained winds of hurricane force, along with a significant storm surge) are possible within 36 hours. A hurricane warning means that hurricane conditions are expected within 24 hours.

Effects of a Hurricane

At the center of a hurricane, the atmospheric pressure can drop as much as 10% below normal. The ocean surface, relieved of pressure in the center of the storm, rises, creating a "dome of water." The rise is greatest in and near the center of the storm, where the barometric pressure is lowest (Fig. 6-5).

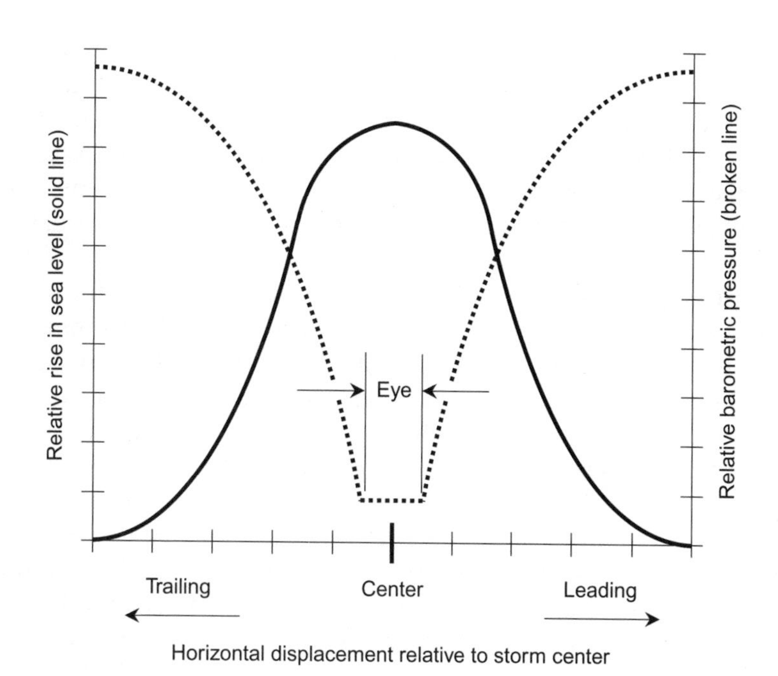

Fig. 6-5. Simplified graph of atmospheric pressure vs. rise in sea level in a hurricane over the open ocean. (This graph assumes the storm is stationary. Forward motion distorts the curves somewhat.)

STORM SURGE

In the open sea, the low-pressure-induced rise in sea level is rarely more than 2 m (about 6 ft). As the center of the hurricane moves into shallow water and approaches a land mass, however, the "dome of water" encounters friction, which exaggerates the rise in the water. The larger the land mass, in general, the more pronounced the effect. The geography of the coastline, the slope of the sea floor near the coastline, and the angle with respect to the coastline at which the hurricane approaches also play a role. The increased friction of the water against the ocean floor causes the water to "pile up," just as it does when incoming swells form breakers on the beach, except more slowly. The power of the hurricane winds, the forward motion of the storm, and the force of the water being driven ashore combine to produce a tide that can rise several meters above normal. The French have an expression for this phenomenon: *raz de maree* (meaning "rise of the sea"). Americans call it the *storm surge.* It is the deadliest phenomenon that occurs in association with hurricanes.

The greater the intensity of a hurricane, the higher is the tide at the center. Onshore winds create a tide by themselves, independent of the fall in the atmospheric pressure. The water is literally pushed up onto the beach. Where the winds blow offshore, the water level can drop, grounding ships in harbors and stranding fish in pools hundreds of meters from the usual shoreline. The most dramatic effect is in the eye itself, where both the wind and the partial vacuum conspire to lift the water.

The storm surge associated with a hurricane may be gradual; the tide simply gets higher and higher as the eye approaches. In some storms, however, the major part of the surge takes place suddenly, in the form of one or more huge, breaking waves, resembling a tsunami. This sudden *raz de maree* is the result of resonance effects in the water, something like the oscillations you can set up in a bathtub. Stories have been told about 10-m (33-ft) waves sweeping in from bays, carrying debris with them, and causing massive destruction within seconds. The worst horror tales come from the Bay of Bengal and the China Sea, where vast populations have settled in low-lying areas adjacent to estuaries and other inlets. A single storm surge in an 18th century hurricane killed 300,000 people in the Ganges River delta area. A similar disaster occurred again in 1970.

The Gulf Coast of the United States, with its shoreline irregularities, is particularly vulnerable to large storm surges. A hurricane might move ashore at a certain place and cause very little *raz de maree,* but if the storm turns and strikes just a few kilometers down the coast, there will be a tremendous surge. The importance of accurate landfall prediction is clear. The Tampa–St. Petersburg bay area, on the western coast of Florida, provides a good example. A storm that moves in from the west and strikes land just south of Tampa Bay will cause no storm surge, and in fact will drive water out of the bay. If the hurricane makes a direct hit, however, or if the eye comes ashore just to the north of the bay, a storm surge of 6 to 8 m (about 20 to 25 ft) is possible. In addition to this, there will be strong tidal currents and large, battering waves.

A storm surge can come with destructive force even outside of an enclosed bay. The famous Miami hurricane of 1926 produced a significant storm surge in the beach area outside of Biscayne Bay. A boat with a 2-m (7-ft) draft completely crossed Key Biscayne during that hurricane. In 1992, Hurricane Andrew caused a 5-m (16-ft) storm surge along the exposed shore near Cutler Ridge, Florida, just south of Miami.

The *raz de maree* can sometimes remove vessels from the ocean altogether. This occurred with Hurricane Camille, in August 1969 when it struck at Gulfport, Mississippi. Three freighters were washed inland, where they ran aground. When Camille departed, the ships were on dry land, standing upright as if they had been moved there by the Army Corps of Engineers!

DEBBY APPROACHES

In 1926, Miami was a small town compared to its present size. The storm of September 18, 1926, would have caused much greater ruin if it had waited for a few decades. It was roughly the size and intensity of Andrew that struck in 1992, except that the 1926 storm came ashore in downtown Miami. Debby is not waiting, and she is worse than either Andrew or the 1926 hurricane.

At 6:00 A.M. Eastern Daylight Time (EDT) on September 16, a hurricane watch is posted for the coastline from Fort Myers to Fort Pierce, Florida. Debby is a definite threat, and she might strike within 36 hours. Residents and vacationers in Miami listen to the advisories concerning Debby. By 6:00 P.M., Debby is strengthening over open water, her eye just to the north of eastern Cuba. The coordinates are given: 22.5° N, 76.4° W. Thousands of pencils each make dots on gridded maps of the Atlantic, Gulf, and Caribbean. Such a map, created especially for following the progress of hurricanes, is known as a *hurricane tracking chart* (Fig. 6-6).

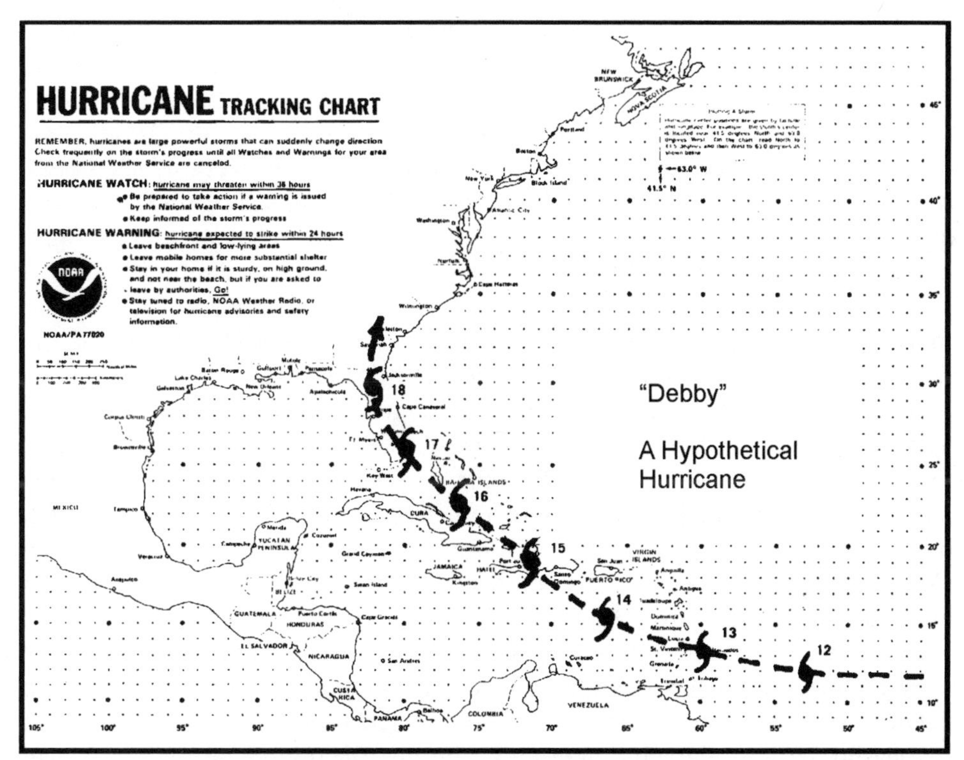

Fig. 6-6. A hurricane tracking chart showing the path of a hypothetical hurricane. The numbers represent dates in September; the time in each case is 6:00 P.M. Eastern Daylight Time (EDT) or 2200 hours Coordinated Universal Time (UTC).

It appears that southeastern Florida lies in Debby's path. Late in the day on September 16, a hurricane warning is issued for the Florida coast from Fort Myers to Fort Pierce. Landfall is expected to occur within 24 hours. Beach dwellers pay special attention to hurricane watches and warnings and evacuate, if necessary, early enough to avoid being cut off from the mainland. Debby threatens to inundate the beach areas with 3 to 5 m (10 to 16 ft) of water. An elderly Miami woman who has seen a few bad storms says, as she packs her car with food and other essentials, "There's only one thing to do in a hurricane. Leave!" The freeways are starting to get crowded on the morning of September 16. By evening, the traffic has become heavy as the warning is issued. Some people are staying, but they are boarding up their windows. The noises of pounding nails and whining electric saws fill the air.

RAINFALL

Hurricanes commonly produce large amounts of rainfall, which can compound the flooding problem and extend it far inland. In times of drought, hurricane rains are welcomed, but if they are extreme, they produce flash flooding because the land cannot absorb so much water in such a short time.

Hurricanes vary in terms of total rainfall. Some storms are "dry," while others are "wet." The most intense hurricanes are not always the wettest, and a weak tropical storm is not necessarily a "dry" one. In 1981, tropical storm Dennis, not a great threat in terms of the storm surge or wind, dropped 51 cm (20 in) of rain in western Dade County, Florida. Although the area had been suffering from a drought, severe flooding occurred. Two years earlier, Hurricane David, a far more powerful storm, hardly wet the sidewalks in much of Miami, despite the fact that his eyewall grazed Key Biscayne and Miami Beach.

When a hurricane strikes land, especially irregular terrain, the winds encounter a sudden increase in friction against the surface. The result can be increased rainfall as the storm spends its energy. If a hurricane slows down or stalls over a certain place, there will be much more rain than if the storm sweeps through quickly. If all of the factors—a "wet" storm, hilly or mountainous terrain, and a slow forward storm speed—conspire together, the rainfall over a period of several days may be fantastic. In July 1913, certain parts of Taiwan (then called Formosa) received about 210 cm (7 ft) of rain from a single tropical cyclone.

In a typical hurricane, the total rain accumulation at any single place ranges between about 10 cm (4 in) and 40 cm (16 in). The total rainfall can be difficult to measure because the wind interferes with the functioning of the rain gauge. During the periods of heaviest rainfall, the wind may blow the rain horizontally, so that only a fraction of the water is caught by the apparatus.

In the United States, the most severe flooding from hurricanes has historically occurred in New England. A notable example is Diane of 1955, which caused over $1,000,000,000 in damage (in 1955 dollars), mostly from flooding. This storm was preceded by several days of heavy rain in Connecticut, Massachusetts, Vermont, and New Hampshire. When Diane came along, the land was saturated and could not handle the runoff. Record 24-hour rainfalls occurred in several places; rivers rose far above flood stage. Another hurricane that caused great flood damage was Hazel of 1954. Hazel moved inland over the Carolina coast of the United States, causing winds of 130 kt (150 mi/h) and massive destruction from beach erosion. As Hazel continued northward, she combined with a low-pressure system from the temperate latitudes, and dropped large amounts of rain in the Great Lakes states. Hazel maintained considerable wind strength for hundreds of kilometers inland. Hazel moved over Toronto, Ontario on October 15 and 16, 1954. Her rainfall, combined with the autumn rains typical of the Great Lakes area, resulted in over $100,000,000 in damage from flooding in metropolitan Toronto alone.

Flooding causes loss of property, as anyone who has been affected by a rampaging river will attest, but flooding can threaten lives, as well. This is more true in the less developed countries of the world than in the United States. Flooding can cause contamination of the drinking water and result in widespread sanitation problems and disease, with which the medical community in an underdeveloped nation cannot deal. Even in a major United States city, the water supply may be made unsafe. Flood waters usually carry debris, such as uprooted trees, automobiles, and fragments of washed-out structures, worsening the damage to buildings and causing numerous injuries to people. Fallen utility lines dangling in the water present an electrocution hazard.

WAVE ACTION

The storm surge and the potentially heavy rains are not the only water-related destructive factors associated with hurricanes. Even if there were no rise of the sea and no rainfall whatsoever, the wave action would still cause devastation immediately along the unsheltered beachfront.

Waves form when wind moves over a water surface, creating friction between the air and the water. We have all observed sizable waves, perhaps as high as 1.5 m (5 ft), on large lakes when the weather is especially windy. In the vast ocean, under the turbulent conditions of a severe and large hurricane, the waves can reach heights in excess of 20 m (66 ft).

When a hurricane is still far offshore, the swells from the storm cause large breaking waves on the beachfront. At a distance from the center of the hurricane, the swells appear to emanate radially outward from the eye (Fig. 6-7). They are actually generated, however, by the rotating winds in the eyewall. The height of a swell depends on the intensity of the hurricane, the diameter of the eyewall, and the distance of the storm from an observer. For beach dwellers, the earliest sign of an approaching hurricane is often an increase in the size of the breakers, and a corresponding decrease in their frequency. The change can be noticed, in some instances, 48 hours prior to the arrival of stormy weather.

The *period* of a train of incoming swells is directly related to their height. Under normal conditions on an ocean beach, the *swell period* is 5 s to 6 s; this translates to a *swell frequency* of 10/min to 12/min. As a hurricane approaches, this frequency drops to around 8/min. As the swell period increases, so does the height. This can produce spectacular breakers. In some parts of the country, surfers flock to the beach to take advantage of these breakers. Unfortunately, the breakers are often accompanied by dangerous *littoral currents* (sideways movement of the water), *rip currents* (large eddies that can carry a swimmer away from the shore in certain places), and *undertow* (a current near the bottom that runs away from shore beneath breakers). All of these phenomena can be dangerous, as can the sheer size of the waves and their uncharacteristic violence. Even well-seasoned swimmers and surfers in excellent physical condition can get into trouble in prehurricane breakers. One of the most horrifying possibili-

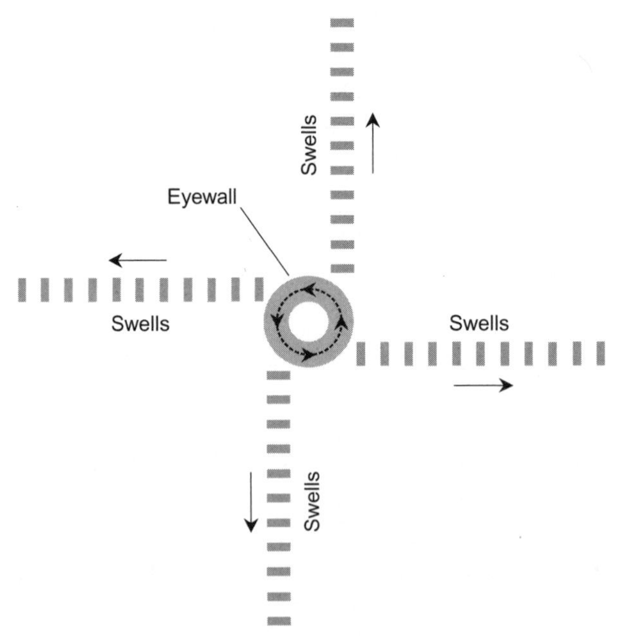

Fig. 6-7. The generation of swells around the eyewall of a hurricane. (This is a simplified rendition that neglects the forward motion of the storm.)

ties is being hurled high into the air after "wiping out" and then falling head-first into shallow water. The result can be a broken neck and permanent paralysis, assuming the surfer does not drown in the turbulent water.

As a severe hurricane nears, the swell frequency can drop to 6/min or less. The thundering of the breakers can sometimes be heard for hundreds of meters inland. There is also a shift in the wind. In the northern hemisphere, the wind backs (shifts counterclockwise as viewed from above) about 90° as the storm nears, so that as you stand directly facing the oncoming breakers, the wind blows against the left side of your body. In the southern hemisphere, the wind turns clockwise about 90°; as you look at the oncoming waves, the wind blows from your right. As the waves break, the wind blows the spray laterally through the air. When a hurricane moves on a track more or less directly toward you, the waves become continually larger, and the direction from which they come does not change significantly. If the storm is not moving directly toward you, the direction from which the waves arrive slowly rotates clockwise or counter-clockwise, depending on whether the eye is traveling toward your right or toward your left as you face the sea.

Ocean swells do not always arrive perpendicular to the beachfront. They can come in at a considerable angle. This is when littoral and rip currents are most likely to arise. If the swells are large, beach erosion occurs. Such swells can wash away much of the shoreline within a day or two. Houses are damaged, roads undermined, and inlets filled in with sand.

TORNADOES AND MICROBURSTS

The wind damage from a hurricane results mainly from the prolonged, violent "blow" in the eyewall. Few hurricanes retain deadly wind speeds farther than about 160 km (100 mi) inland, although this depends to a large extent on the terrain, and considerable damage can still result from sustained winds of 50 to 60 kt (about 60 to 70 mi/h). Hurricanes can produce tornadoes, however, and these have higher wind speeds than those in the general circulation of the eyewall.

Tornadoes are most frequently observed in the forward semicircle of a hurricane, especially in the right front quadrant. But they can take place anywhere in the rainbands or eyewall the storm. In a hurricane, a tornado may move much faster than a typical "Texas twister," and can strike with little or no warning. The tornadoes associated with a tropical hurricane are, however, rarely as large or violent as their continental counterparts. Hurricane-spawned tornadoes can nevertheless tear the roofs off of buildings, shatter windows, overturn automobiles, strip trees, and cause other serious damage to property. As the hurricane moves into the temperate latitudes, the tornadoes may become more violent. Hurricane Camille, in August, 1969, caused significant tornado damage throughout the Gulf Coast and southeastern United States as she moved northeastward past the 30th parallel.

Microbursts can also occur in hurricanes, especially in the eyewall where the most intense thunderstorm activity is found. These are the same phenomena sometimes seen in strong thunderstorms associated with frontal cyclones in the temperate latitudes. As the air descends and strikes the surface, it "spills out" in all directions like a water stream striking a hard floor, momentarily adding to the wind speed in some directions, momentarily reducing it in other directions, and momentarily shifting its direction in some places.

An exceptionally strong wind gust or lull in the eyewall of a hurricane, particularly if it is attended by a perceived change in the air pressure (felt on the eardrums), can signify a passing tornado or microburst. The author noticed several instances of this phenomenon during the passage of the forward portion of the eyewall of Hurricane Andrew in Homestead, Florida. The wind would suddenly abate for a couple of seconds, and high pressure was felt on the eardrums. After the storm, some trees were blown down at angles much different from what would have been expected if the winds had blown in straight lines at all times. This lends support to the hypothesis that the eyewall of Andrew contained tornadoes or microbursts, or both.

LIGHTNING

The amount of electrical activity in a hurricane can vary from practically zero to a dazzling, continuous display. Some severe hurricanes have little or no lightning. As Hurricane David passed by Miami, in 1979, little or no sferics ("static") was observed at shortwave and medium-wave radio frequencies. But tropical storm Dennis of 1981, a much less intense storm, produced lightning so brilliant and continuous that it was possible to read a book by its light during the nighttime. Dennis was less organized, but produced much more rainfall, than David.

Thunder is easier to hear in the less violent part of a hurricane than in the eyewall, and lightning is more readily seen at night than during the daytime. With a high wind, people are not likely to notice (or care about) lightning and thunder. Heavy rains and limited visibility can also make lightning hard to detect. When lightning is seen in an intense hurricane, it often has strange colors such as yellow, red, or green.

THE EXTRATROPICAL HURRICANE

The hurricane of the tropics is a symmetrical storm, having isobars that form almost perfect circles. It is somewhat more intense, and covers a slightly wider area on the side in which the wind blows in the same direction as the storm moves. In the northern hemisphere, this is the righthand half of the whirlwind; in the southern hemisphere, it is the lefthand half. There are no frontal systems associated with a tropical hurricane.

As a hurricane travels farther from the equator, it gets less symmetrical, develops warm and cold fronts, and attains a higher forward speed. The latitude at which a hurricane changes in character from tropical to extratropical, or the exact moment the transition takes place, varies from storm to storm. It makes a difference whether it is early in the hurricane season, at the height of the season, or late in the season. Between approximately August 15 and October 15, hurricanes are most likely to retain their tropical characteristics some distance into the temperate zone. Nonrecurving storms do not become extratropical. They remain in the tropics, and either die off at sea or over land.

Most extratropical hurricanes usually have less violent winds than tropical hurricanes, but there are exceptions, such as the famous 1938 hurricane that struck Long Island and New England. The winds blow much harder on the side of an extratropical storm in which the forward movement adds to the wind circulation. Some extratropical hurricanes move forward at speeds as great as 40 to 50 kt (47 to 58 mi/h). This can make a big difference in surface wind speed on the "strong" side of the storm as compared with the "weak" side.

Extratropical hurricanes have identifiable eyes less often than tropical hurricanes. The pressure gradient near the core is less pronounced. The eyewall becomes irregular or elongated, and the eye fills with clouds. Warm fronts and cold fronts form. Thunderstorms and squall lines occur, especially in association with cold fronts over land. Heavy rain and high winds are often observed hundreds of kilometers from the center. The eyewall eventually disappears altogether. If a late-season hurricane wanders far enough from the tropics, the rain changes to snow in some sections of the storm.

NAMING OF STORMS

As the hypothetical Hurricane Debby bears down on Miami, the residents might wonder how this benign-sounding name can reasonably be used as the code word for such a monstrous storm. Obviously, the name has little to do with the character of the storm. Why do we name hurricanes at all?

During the Second World War, hurricanes—especially western Pacific typhoons—caused trouble for the United States Navy. Often, more than one storm was in progress at a given time, and all of them were constantly moving in various directions. Confusion arose as to which storm was which. There were no weather satellites back then, and it was more difficult to locate the center of a storm than is the case today. The military started to name hurricanes to keep track of them and to tell them apart. At first, the standard military phonetics for the alphabet were suggested as possible names for hurricanes. Thus, the first storm of a season would be named Able, the second one Baker, the third one Charlie, and so forth. (Military phonetics have changed since that time, so if this scheme were adopted today, the first three storms would have been called Alpha, Bravo, and Charlie.)

The main problem with the phonetic scheme soon became evident: Which *ocean* was which? There might be three hurricanes called Baker: one in the Caribbean Sea, one in the Pacific Ocean, and one in the Indian Ocean. While the idea of getting one ocean confused with another seems silly at first thought, such mix-ups can and do happen. It wouldn't be very funny when a destroyer got caught in a storm such as Debby because the captain thought it was in the eastern Pacific Ocean and not in the North Atlantic! In order to prevent this sort of problem, separate name lists were composed for each oceanic region (North Atlantic, Eastern North Pacific, Central North Pacific, Western North Pacific, and so forth) in each hurricane season.

The naming of hurricanes was not formally considered until after the Second World War. In 1953, the official phonetics were changed, and a two-year plan for giving hurricanes phonetic names was dropped. It was decided that female names would be used instead. Different lists of names would be used for different oceans to eliminate the possibility of the sort of confusion just described. In the next year, 1954, hurricanes Carol, Edna, and Hazel raked New England, the middle Atlantic states, the Ohio Valley, and the Great Lakes region. In 1978, it was decided that hurricanes should receive male names as well as female names. David of 1979 became the first big "male hurricane."

Developing tropical cyclones are not named until they attain tropical storm intensity by having sustained winds of 34 kt (39 mi/h) or more. Before then, the systems are sometimes given numbers identifying them in order of occurrence for the season, for example, "Tropical Depression Five." If a hurricane or tropical storm loses its intensity and returns to tropical depression or disturbance status, it loses its name. If it regains strength and once again acquires gale-force or hurricane-force winds, it gets its old name back.

Preparation and Survival

For people who live in hurricane-prone regions, it is wise to make some arrangements for a possible storm at the beginning of each hurricane season. In North

America, the season officially starts on June 1. Nonperishable foods, tape, flashlight cells, and other storable items should be obtained at that time. When watches and warnings are issued, there will likely be spot shortages of critical supplies.

WATER

It is especially important that an adequate water supply be maintained. After a hurricane has passed through a region, municipal water can become contaminated and unsafe. *Saltwater intrusion* can occur as a result of the storm surge. Sewage can sometimes become mixed with water. Flooding can cause trash, fertilizer, and animal waste to work their way into the drinking water mains.

If a hurricane warning is issued, the best way to store water is to fill every available container before the storm hits. The big plastic bottles in which milk is sold are ideal for this purpose. Bottled water can be purchased, as well. All sinks and bathtubs should be thoroughly cleaned, rinsed, sterilized with chlorine bleach, and the drains made completely watertight using a non-water-soluble, non-toxic sealant. Such vessels should be filled with water. A couple of drops of chlorine bleach per liter (approximately 1 quart) of water can prevent harmful bacteria from growing for several days.

FOOD AND COOKING

A substantial store of nonperishable food is important, as well. This can include canned meats, vegetables, juices, fruits, nuts, and canned or evaporated milk. Such foods should be purchased before the start of the hurricane season, and can be used up slowly after the end of the season. Don't forget to buy manual (non-electric) can openers! After a hurricane, the electricity might be off for a long time, and electric can openers won't work. If it appears possible that a storm might strike within several days, certain foods can be purchased that will keep for a few weeks in a functioning refrigerator, or for a few days in a closed refrigerator without electricity. Examples of such foods include certain processed cheeses, fresh fruits and vegetables, bread, and hard-boiled eggs.

It is best to get a good supply of foods that can be eaten without cooking. If it is necessary to cook foods, and if you happen to have a gas stove or oven with your own independent gas tank, be sure the tank is filled a long time before a storm looms. (Don't try to call the service people after a hurricane watch or warning has been issued!) During the storm, the tank valve should be closed, and no attempt should be made to cook anything. Wait until the hurricane has passed. Charcoal should never be used indoors, because the combustion produces deadly carbon monoxide gas which is colorless and odorless. Portable burners, intended for camp use, are usually alright to use indoors, but be careful with fire and be sure to read all the instructions before you use them. Be sure you have plenty of fire extinguishers around, and be sure they have been maintained so they will work if you need them.

OTHER STEPS TO TAKE

Fill the gas tanks of all your vehicles. After the hurricane, station pumps may not work because they use electric motors. If you have a boat, it should be moved away from the ocean or bay. Don't wait until the storm is almost upon you; get it to safety a day or two in advance of the hurricane. Boats on trailers should be kept indoors if possible. If that is not possible, the boat and the trailer should be anchored down with heavy aircraft cable and clamps—although a severe hurricane may have winds strong enough to snap aircraft cables or rip the anchors out of the ground, and carry a boat hundreds or thousands of meters away.

If time is available, prune down trees and shrubbery to protect them against destruction by the wind, and to reduce the danger of damage from flying limbs. (Ideally, pruning should be done, and kept up to date, on a routine basis.) Even a minimal hurricane can uproot large trees. Take the cut branches to the dump; do not leave them lying around. If you have coconut trees, have someone take or knock down the coconuts. A hurricane wind can dislodge them and propel them through the air like cannon balls.

The winds of a hurricane can shatter windows as a result of air pressure. Flying debris compounds this problem. Windows should be boarded up with the thickest plywood available, using plenty of large nails. Alternatively, windows can be protected by metal shutters. If this is not feasible, masking tape should be placed in an X-pattern over the inside and outside of every window pane. Of course, once the storm strikes, you should stay as far as possible from windows, no matter how well secured they are.

Loose objects can become lethal missiles in a hurricane wind. Trash cans, radio antennas, bicycles, tools, lawn mowers, and other objects can be hurled for hundreds of meters at high speed. Such objects must either be stored indoors or tied down with aircraft cable.

DURING THE STORM

False rumors will circulate during the days and hours before a hurricane strikes. It is important to get the facts. The National Hurricane Center will disseminate the information over local radio and television stations. If you live near the beach or bay shore, and you are advised to evacuate to higher ground, do it!

As the eyewall of the storm moves inland, electricity will be lost. A battery-powered radio, with two or three sets of spare batteries (that have been checked and found good), should be available for use during the blackout. A good flashlight, with several sets of spare batteries and one or two spare bulbs, is a necessity as well. The lantern-type light, with the large battery, is excellent. Candles and matches can be used if nothing else is available, but they create a fire hazard. Candles molded in low, wide glass jars are least likely to be tipped over. When power is lost, switch off all lamps and appliances to protect them against a voltage surge when power is restored. Shut off the main switch or switches at the distribution box.

The hours immediately before the arrival of the eyewall may be deceptively calm. The rain and wind may diminish to a drizzly breeze. But when the eyewall arrives, there will be no doubt about it. The wind will kick up and things will begin to fly through the air. Move to the leeward side of the building (the side facing away from the wind). Children and elderly people will need reassurance, because the sound of the wind can be traumatizing. Animals may have to be placed in restraining boxes or crates.

THE SAFFIR-SIMPSON SCALE

Hurricanes vary greatly in size, forward speed, and circulating wind speed. The circulating wind speed is inversely proportional to the barometric pressure in the center. The storms with the lowest pressures in their eyes invariably have the greatest circulating wind speeds. The potential storm surge increases with the circulating wind speed, but is also affected by the forward speed of the storm, the diameter of the eyewall, the angle with which the hurricane approaches the coast, the contour of the sea floor near the coast, and the general shape of the coastline in the path of the storm.

A numerical scale has been developed for categorizing hurricanes in terms of their severity. This is known as the *Saffir-Simpson scale,* and it is outlined in Table 6-1. Storms of category 3 or more are considered major hurricanes. Category 5 hurricanes do not occur very often, but when one of them strikes a populated area, the result is always a catastrophe.

PROBLEM 6-5

If the eyewall of a strong hurricane passes over a certain place, does this invariably mean that there is a massive storm surge at that location?

Table 6-1. Saffir-Simpson intensity scale for hurricanes. Wind speeds are sustained (based on a 1-minute average.)

Saffir-Simpson category	Wind speed in knots	General observed effects
1	64–82	Minimal hurricane. Some trees blow down, mobile homes overturn, unprotected windows break or blow out. Storm surge of 2 m (6 ft) or less occurs.
2	83–95	Moderate hurricane. Roofs blow off houses. Major damage is done to beachfront properties. Numerous trees blow down. Storm surge of 2 to 3 m (6 to 10 ft) occurs.
3	96–113	Major hurricane. Many houses and buildings on the beachfront are destroyed. There is considerable damage to houses inland. Few trees are left standing. Storm surge of 3 to 4 m (10 to 13 ft) occurs.
4	114–135	Severe hurricane. Substantial damage to property occurs inland, and almost total destruction is observed on the beachfront. Landscape defoliation is extensive. Storm surge of 4 to 5.5 m (13 to 18 ft) occurs.
5	136 or more	Extreme hurricane. Total destruction takes place along the beachfront. Significant structural damage is done to most buildings inland. Landscape is almost totally defoliated. Storm surge of 5.5 m (18 ft) or more occurs.

SOLUTION 6-5

No. If the lefthand portion of the eyewall (in a northern-hemisphere storm) or the righthand portion (in a southern-hemisphere storm) passes over a point on the coast, the tide can drop below normal. This is because the high winds blow offshore, pushing water away from the coast. However, depending on the shape of the coastline in the region, there may still be some storm surge before or after the eyewall passes over.

Debby Strikes

As the warning is issued, the broadcasts are the only visible evidence of what is coming. The sky has a few high cirrus clouds, with scattered cumulus at lower levels. The sun sets with the ruddy glow familiar to Floridians. The barometer has dropped slightly, but not yet to an unusual extent. The large ocean swells, thrown out in advance of the storm, do not get to the beaches because of the reefs offshore.

Later, cumulus clouds move more quickly than usual across the sky, and they assume a dusky gray appearance. In terms of diameter, Debby is not a particularly large hurricane, and she is still about 725 km (450 mi) away, approaching at a little less than 18 kt (20 mi/h).

SEPTEMBER 16, 9:00 P.M. EDT

Debby appears to be headed straight for Miami, with maximum sustained winds of 145 kt (167 mi/h). The director of the National Hurricane Center tries to describe the fury of Debby to the residents. Some of them remember previous hurricanes, and need no description.

A radar picture, obtained from a reconnaissance aircraft near the eyewall, is shown on television. The eye of the storm shows up clearly, as do the eyewall and the spiral rainbands. Preparations for the arrival of Debby continue throughout the night. Most people are serious about this; they are busy boarding up windows, securing loose objects, and taking other emergency measures. Hardware and grocery stores remain open until all their supplies are gone. Gas stations, too, remain open.

SEPTEMBER 17, 2:00 A.M. EDT

The full moon is suddenly obscured by the cumulonimbus clouds of the outermost rain band of the hurricane. The wind gusts fitfully from the northeast. The shower is over within a few minutes, and the moon reappears dimly behind layers of fast-moving clouds. Every few moments a ragged fractocumulus or "scud" cloud temporarily blocks the moon; then the silvery disk reappears.

The full moon means that the normal gravitational tides will be at their highest, and if the hurricane tide strikes along with the occurrence of normal high tide, the storm surge will be extreme. This, say the forecasters, is the main reason for coastal dwellers to evacuate inland. Most heed the warnings; a few do not. The next rain squalls move into Miami at about 4:30 A.M. The smattering of rain against boarded-up windows awakens the residents that have managed to fall asleep despite the growing tension and apprehension. A glance out the door reveals slanting rain, swaying trees, and a more rapidly moving rack of gray clouds, illuminated by the lights of the city. The showers again pass, but the weather grows worse by the hour.

SEPTEMBER 17, AFTER SUNRISE

Rain showers and thunderstorms continue intermittently throughout the morning hours, but none are violent, and some people are getting the idea that Debby might be an overrated storm. The forecasters warn, however, that Debby is an unusually fast mover. The destructive eyewall will strike with greater than usual rapidity. It also means, fortunately, that the storm duration will be shorter than that of the average hurricane, as long as Debby doesn't stall or slow down. The eyewall is small, having contracted to an outside diameter of about 95 km (60 mi). If Debby makes a direct hit, she will get her work done in a hurry. People are talking about this storm as Andrew (of 1992) reincarnated, and they fear Debby will be worse because she appears to be on track to pass over downtown Miami and Miami Beach.

SEPTEMBER 17, AFTERNOON

Just after noon, the sky lightens a little, and for a few minutes the animated trees cast their shadows on the damp ground. The wind along the beaches is now a solid northeasterly gale. Inland, conditions resemble any squally, windy day in September. But the barometer is falling fast, and this unusual phenomenon is noticed even by people who don't know anything about hurricanes.

Coral Gables, a southwestern suburb of Miami, is graced with green, lush foliage and beautiful homes. Although the residents don't yet know it (nor can the forecasters precisely predict it at this point), Coral Gables lies squarely in the path of Debby's core. Trees begin to fall at about 3:00 P.M. as the innermost rainband passes through the area, accompanied by two tornadoes, driving rains, lightning, thunder, and wind gusts to 80 kt (92 mi/h). The power goes out. Broken utility lines flap and spark in the gale.

Just before 4:00 P.M., a milky green, sideways-moving wall of rain and debris swallows up the landscape from the southeast.

THE SIEGE: PHASE 1

As the edge of the eyewall approaches, the sound is like that of a freight train passing. The house in which you sit begins to shake, even though its exterior walls are made of solid concrete. You can hear things smacking against those outer walls and against the plywood covering the windows. The air becomes filled with tree limbs and whole palm tree tops, unsecured bicycles, lawn mowers, boats, Spanish tiles, broken glass, and big signs. Automobiles are over-

turned, and some smaller cars are rolled until they come to an obstruction. The rain lashes with the force of a firehose. Corrosive salt spray from Biscayne Bay, blown for miles inland, mixes with the rain. Visibility is nil.

At the National Hurricane Center, the anemometer is destroyed shortly after the official sustained wind speed exceeds 130 kt (150 mi/h). The true wind speed will never be measured. Inside their homes, residents huddle in broom closets, under kitchen and dining room tables, and even under mattresses. For a short time, some radio stations, equipped with emergency generators, continue to operate, giving the latest radar information and other data. One by one, the broadcast stations go off the air as their towers collapse.

The blow continues unabated. Along the bayfront and beach, the onshore winds push the water higher and higher. As the edge of the eyewall moves up the coast, the few people who have remained on the waterfront discover why they were advised to evacuate. The sea threatens to carry their homes away and then drown them outright. Windows in highrise buildings blow out, followed by the destruction of the inside walls. People lie flat on the floors. The wind tide around the northern periphery of the eyewall reaches 6 m (20 ft) above normal in some locations, washing sand and silt into the lower stories of the buildings. Some houses and condominiums collapse as 3-m (10-ft) waves undermine their foundations. Miami Beach and Key Biscayne become part of the ocean floor for two long hours. When the water subsides, most of the causeways will be gone. The only transportation to the mainland will be by boat. The beach will have been rearranged. Sand will be knee deep on Collins Avenue.

THE EYE ARRIVES

Finally, at about 5:00 p.m, some of the coastal dwellers notice that the roar of the wind is lessening. A break is coming. The visibility improves and the sky lightens. The rain slackens, and then stops altogether. The edge of Debby's eye is moving ashore. The eye of this storm is small and well-defined. The small size of the eye, combined with the fact that Debby is moving fast, means that the lull in the center will not last long. Those who lie precisely in the path of the storm will have less than half an hour of respite before the rear semicircle of the eyewall strikes. In most places where any lull is observed, it will last for a much shorter time. Most of the residents will not experience the passage of the eye, but will notice a rapid shift in wind direction, veering in the righthand half of the storm and backing in the lefthand half.

Some people cautiously peer out of their battered homes at the destruction around them, taking advantage of the momentary letup in the wind and rain. Others, only a few kilometers away, are still experiencing the full fury of the hurricane.

The clouds in the eye are thin and broken. The blue sky is visible in patches. The slanting, late-afternoon sun illuminates the towering banks of cloud. The shadow of the western eyewall falls on the eastern eyewall. The center of Debby is a "hole in the sky" 13 km (8 mi) across and equally deep. All of the clouds are continually in motion. The roar of the wind can be faintly heard, like distant aircraft circling.

The air within the eye of a hurricane is usually warmer than the air surrounding the eye. A hurricane is a warm low-pressure system, and the humidity is often observed to fall in the core of a storm. Such is the case with Debby. The barometer on the wall in the living room shows 27.05 inHg. Normal atmospheric pressure will support about 30 inHg. Thus, the pressure within the eye of Debby is only 90% of normal.

THE SIEGE: PHASE 2

The sky begins to darken toward the south and east; the wind has shifted to the southwest. Rain begins to fall again. Then the rear half of the storm moves in, and chaos returns. After the period of light winds and absence of rain, the storm seems even more violent than it was before the arrival of the eye. In some hurricanes, the second half is actually more violent than the first. This is especially common in storms that pass over islands. In the case of Debby, however, the second half is slightly weaker in Miami, because the wind no longer comes directly off the ocean.

A few people get caught outside as the storm returns, taken by surprise at the suddenness with which the eyewall has moved back over them. One man struggles and slides on the wet concrete of his driveway. His wife throws him a piece of clothesline and pulls him in. During the next hour and a half, what little is left of the foliage in Coral Gables is stripped of almost every leaf. Some of the stronger pine trees remain standing; others lean as if they have been pushed down by a great bulldozer. Houses and apartment buildings are unroofed, and the rain comes rushing in.

By 7:00 P.M., the worst of the storm has passed. Nightfall arrives with a deepening gray.

AFTERMATH

The full extent of the damage from Debby will not be realized for some time. All telephone service has been wiped out. There is no electricity. Broadcast stations with antennas that lay in the path of the eyewall are off the air. Along the beachfront, many buildings have been washed away. Only the foundations, some obscured by sand, remain. Fish are strewn everywhere, some with their eyes popped out. Streets and highways have been undermined in some spots, causing the pavement to collapse.

Hurricane Debby has moved into the Florida peninsula, and her violence has decreased. Orlando is buffeted by squalls and tornadoes. The storm drenches the swamplands to the west of Jacksonville, and moves on into southern Georgia. Heavy rains cause flooding throughout the southeastern United States. Then the remains of Debby combine with a temperate low over Pennsylvania, producing floods in the Northeast.

Certain precautions must be observed following a hurricane. Even if power is out, some fallen wires may carry thousands of volts. If live wires are dangling or lying in a puddle of water, people can be electrocuted even without coming into direct contact with the wires. Stay away from standing water and downed power lines.

The drinking water, if not shut off by the municipal authorities, cannot be guaranteed safe for human consumption. As broadcast stations return to the air, advisories will be given concerning the integrity of the water supply. Do not use the water until you are told it is safe to do so by a reliable source.

If your telephone still works, don't try to use it except in an emergency. The switching networks, if they function at all, will be working at only part of their normal capacity, and the volume of attempted calls would overwhelm them even if they were functioning at their best.

Do not attempt to drive your car or truck (or even a Land Rover) until most of the debris has been cleared. After a storm such as our hypothetical Debby, you might run over debris and get a flat tire, get stuck in sand or mud, or get lost because of the destruction of street signs and landmarks.

PROBLEM 6-6

Suppose that Debby had been moving at only half the forward speed as was the case in the scenario just described, but all other factors were equal (same intensity, same diameter, same storm track). What difference would this have made?

SOLUTION 6-6

Hurricane conditions at all points would have existed for twice as long. The maximum sustained winds would have been a few knots lower on the strong (right-hand) side of the hurricane and a few knots higher on the weak (left-hand) side. The amount of rainfall would have been approximately twice as great at all locations.

Quiz

This is an "open book" quiz. You may refer to the text in this chapter. A good score is 8 correct. Answers are in the back of the book.

1. Well-defined cold fronts and warm fronts are most likely to be observed in a hurricane if it
   (a) acquires tropical characteristics.
   (b) reaches category 3, 4, or 5 intensity.
   (c) passes from the tropics into the temperate latitudes.
   (d) fails to recurve.

2. The term recurve, with respect to a hurricane, means that a storm
   (a) does a 360° "loop-the-loop."
   (b) follows a path that turns toward the pole.
   (c) follows a path that turns toward the equator.
   (d) follows a straight path in the tropics.

3. The rainfall in a hurricane can be difficult to measure because
   (a) anemometers rarely survive the extreme winds in the eyewall.
   (b) the wind can interfere with the ability of a rain gauge to catch the rain.
   (c) the amount of rain is often greater than any instrument is designed to measure.
   (d) the barometric pressure can get too low for a rain gauge to work properly.

4. As a hurricane approaches and ocean swells or breakers become larger, the frequency of the swells or breakers
   (a) increases.
   (b) stays the same.
   (c) decreases.
   (d) rotates counterclockwise.

5. Suppose you are standing on a beach on the Gulf Coast of southwest Florida and there is a tropical storm offshore, directly west of you. Imagine that you are well within the circulation of the storm, so you are experiencing gale-force winds. These winds blow from the
   (a) south.
   (b) north.
   (c) east.
   (d) west.

6. The lowest pressure in a hurricane is observed in the
   (a) eyewall in the forward semicircle.
   (b) eyewall in the rear semicircle.
   (c) rainbands.
   (d) eye of the storm.

7. When hurricanes were first named after the Second World War, which of the following schemes was initially used?
   (a) American first names.
   (b) Phonetic representations for A, B, C, etc.
   (c) American surnames.
   (d) Spanish nicknames.

8. A tropical depression is given a name, and considered to be a tropical storm, as soon as
   (a) its maximum sustained wind speed reaches 100 kt (115 mi/h) or more.
   (b) its maximum sustained wind speed reaches 64 kt (74 mi/h) or more.
   (c) its maximum sustained wind speed reaches 34 kt (39 mi/h) or more.
   (d) it has caused serious human casualties or property damage.

9. A storm surge can cause
   (a) general flooding.
   (b) buildings and roads to be undermined.
   (c) saltwater intrusion into the water supply.
   (d) All of the above

10. The most severe type of hurricane is
    (a) a temperate frontal cyclone.
    (b) a tropical depression.
    (c) a gale force storm.
    (d) a category 5 storm.

CHAPTER 7

Winter Weather

Where snow falls in abundance, winter is great for skiing; and where cold temperatures freeze lakes deep enough, it is good for ice skating. But winter can also be dangerous. A large winter storm over North America can expend as much energy per unit time as a major tropical hurricane. The Omaha tornado of 1975 killed three people. But in 1984, a blizzard was blamed for more than 20 deaths in the midwestern and northeastern United States.

The Stormiest Season

People who live in the middle latitudes in the southern hemisphere are battered by continental winter storms less frequently than people who live in the middle latitudes in the northern hemisphere. This is because, between the Tropic of Capricorn and the Antarctic Circle, most of the earth's surface is covered by ocean. There just aren't as many people down there! In the northern hemisphere, the continents of Europe, Asia, and North America compose much of the world's total land mass. A sizable proportion of each continent is situated in the prime winter storm track. Cyclonic winter weather systems, spun off by the semipermanent Aleutian and Icelandic low pressure systems, pass over populated regions intermittently between September and May. The most intense of these occur from late November through March, peaking in January or February.

LENGTH AND SEVERITY

At any given place, the length and severity of the winter season depends on two factors: the latitude and the distance from the ocean. As distance from the equator increases, in general, winters get colder and longer. However, a town in the center of a large continent has a longer and more severe winter storm season than a coastal town at the same latitude. (An especially interesting example is a comparison between the climates of southern Italy and the Great Plains of Nebraska, which are on the same line of latitude.) Topography and altitude also have an effect. Mountains influence temperature distributions and wind patterns.

RIDGES AND TROUGHS

The jet stream can change position and contour with amazing swiftness, moving north or south at any given longitude by hundreds of kilometers in a single day. But in some winter seasons, it seems to get "locked in" to a certain mode.

A persistent winter ridge (anticyclonic bend in the jet stream) over the center of North America can produce a warm, dry season with little snow in the United States, but the risk of a subsequent summer drought. A persistent winter trough (cyclonic bend in the jet stream) over the central or south-central United States can result in a cold season with above-average snowfall, high heating bills, and spring flooding. A jet-stream trough along the West Coast can cause heavy rains in southern California and snowy winters in the Sierra Nevada mountains, while a ridge produces a warm, dry winter in the same locations. A persistent trough along the East Coast brings repeated storms to New England and cold fronts to Florida; a persistent ridge in that region brings less precipitation to New England and warmer weather to the vacation havens of Florida.

Predicting jet-stream behavior is almost as difficult as predicting whether a coin will show heads or tails on the next flip. The science of long-term weather forecasting is improving, but is still inexact.

THE GALES OF AUTUMN

Sometime during the autumn, most regions in North America, Europe, and North Asia get a brief warm spell. Most, or all, of the leaves have fallen from deciduous trees, and the days rapidly grow shorter and crisper. There is an intermission that people in America call *Indian summer*. Then one day the wind veers sharply, the sky becomes overcast, and the temperature plummets. Not every winter begins like this, but many do. A strong cyclone, the first in the winter series, pushes through. In Canada, it can happen as early as mid-September. At progressively lower latitudes, the first storm arrives progressively later, perhaps after Thanksgiving, in such places as Oklahoma or Virginia. In the extreme southern part of the country, such as Texas and Louisiana, these rainy gales represent the deepest part of the winter season.

In the Great Lakes and Northeast regions of the United States, November can produce storms as violent as minimal hurricanes. Some of these storms have been immortalized in history books. In the year 1641, a "November gale" battered and flooded parts of New England, causing hurricane-like storm surges and beach erosion. In 1950, a monstrous cyclonic system, called the "Great Easterly Gale," moved through the Ohio Valley and New England. In some locations, sustained winds rose to 65 kt (75 mi/h), and peak gusts reached approximately 90 kt (104 mi/h). Damage to trees and buildings was considerable. The midwestern United States is characteristically hit by westerly winds of gale force on at least one occasion, and in some seasons several times, every autumn. The Pacific Coast is not spared, either. That is where autumn and winter storms in North America make landfall, and their high winds and large waves have been well publicized.

European history contains accounts of gales that have caused widespread damage to forests and buildings. November was known in old English lore as the "windy month." A late-autumn Atlantic storm in the 1700s sent residents of southern England under tables and into closets as their houses shook, windows burst, and debris filled the air. Some homes in the countryside were demolished by the wind. The event was accompanied by storm surges along the western coast, and the system even had a lull followed by a wind shift. Detailed data on this storm are lacking, of course, because it took place before the era of satellites and radar. However, its reputed intensity suggests that it was a late-season hurricane that traveled "off the beaten track."

SLEET AND ICE STORMS

During November, many parts of the northern hemisphere experience their first sleet and freezing rain. The "ice belt" progresses toward the south through December. Sleet and freezing rain are among the most dangerous winter weather phenomena, primarily because they make it dangerous to drive motor vehicles. After the carefree driving of summer and fall, it takes awhile to get used to the change in road conditions that winter inevitably brings.

Sleet resembles small hail. Sleet pellets are about 1.5 mm (1/16 in) to 3 mm (1/8 in) in diameter. In a cloud, water is normally frozen at high altitude. As ice pellets or snowflakes fall, they melt into raindrops as they encounter warm temperatures at the lower altitudes. But if the temperature of the air near the surface is cooler than the air near the cloud base—above freezing but below approximately 4°C (39°F)—the raindrops may refreeze before they reach the surface (Fig. 7-1). In that case, the result is sleet.

Freezing rain occurs for a different reason. The ice pellets melt as they fall, just as they do in an ordinary rain shower. The precipitation arrives at the ground in liquid form, but freezes as soon as it lands. Freezing rain is most likely to fall when the temperature of the surface, and of objects at the surface, is at or below freezing while the temperature of the air is significantly above freezing (Fig. 7-2). In some situations, these conditions can result in a coating of ice in excess of 2.5 cm (1 in) thick on every tree, house, utility wire, automobile, and roadway. But usually, the ice thickness is less than 1 cm (0.4 in). If the icing is significant, the event is called an *ice storm.* A severe ice storm is doubly dangerous, because it not only slickens the roads, but tree branches and utility wires often break and fall. This can block roadways, and falling limbs may strike moving vehicles. In an ice storm, driving should be avoided except in emergencies.

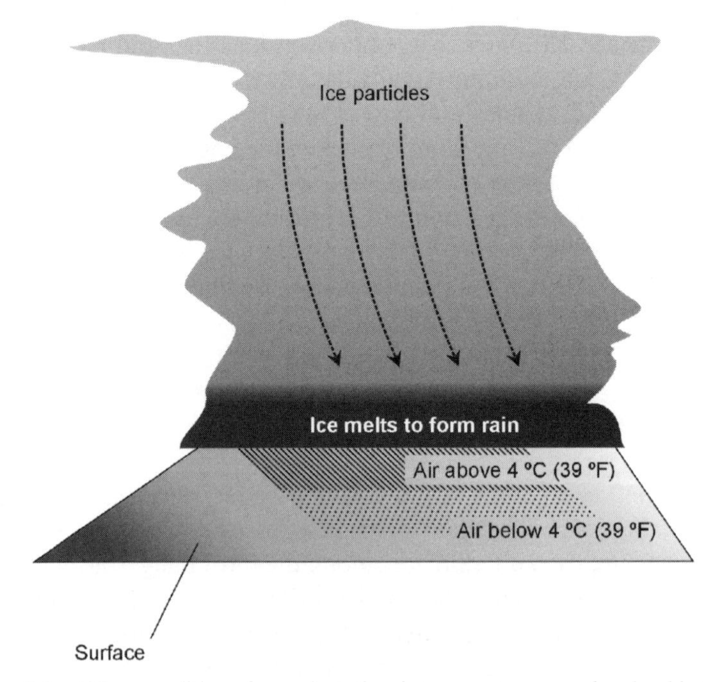

Fig. 7-1. When conditions favor sleet, the air temperature near the cloud base is far enough above freezing to ensure that the precipitation falls from the cloud as liquid, but the temperature at lower altitudes is low enough to cause refreezing before the precipitation reaches the surface.

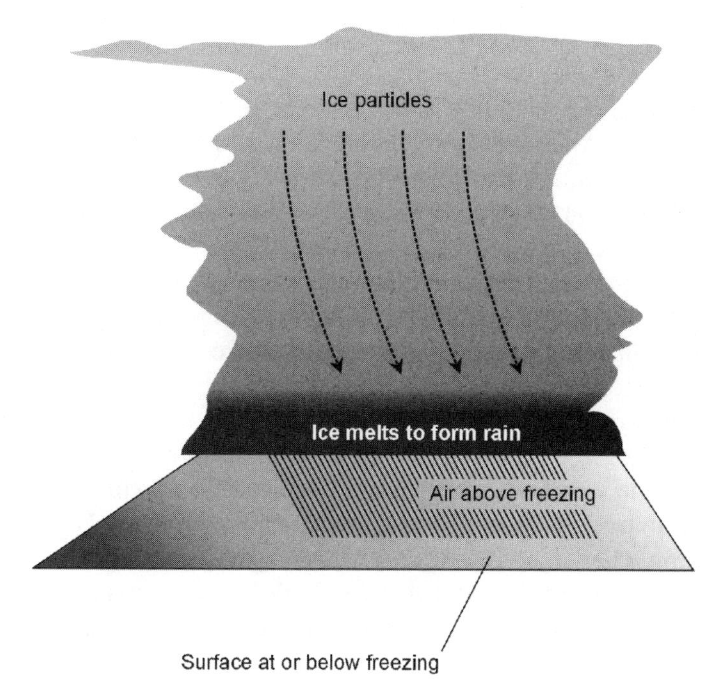

Fig. 7-2. When conditions favor freezing rain, the surface temperature is at or below freezing, while the air temperature remains warm enough to allow precipitation to fall as rain.

In the United States, the southern and eastern regions are most frequently affected by ice storms, although the Midwest gets its share. Ice storms are less common west of the Rockies and in the far north. A severe ice storm, followed by a cold snap, can produce a layer of glaze that lasts for days. In 1978, an ice storm in southern New England kept trees and utility wires coated with glaze for a week. The effects of an ice storm are worsened if snow falls immediately before or after the icing event. An ice storm accompanied or followed by high winds can wreak tremendous damage to trees and utility lines.

Sleet or ice storms sometimes precede snowfalls, and sometimes follow snowfalls. From the point of view of a motor-vehicle driver, it is better if sleet or ice falls on top of the snow rather than underneath the snow. If the ice is on top, the plow gets rid of the ice as well as the snow. Children prefer the ice on top, too: they like to walk and slide on the snow without sinking in! Snow blowers are often foiled by snow with a heavy frosting of ice on top. But for those who shovel the stuff, it doesn't make much difference.

SNOW ZONES

Much of our planet is covered by snow all the time. The highest mountain peaks, Greenland, and the Antarctic continent are almost totally blanketed by snow all year round. In the northern hemisphere, snow falls at least once a year over about 50% of the land. We can draw a line around the globe, representing the demarcation between regions where it normally snows and regions where it doesn't snow very often (neglecting mountain peaks). This is done by defining areas where the average temperature is below freezing during the coldest 30-day period of the year. The longest winters in the civilized world occur in the interiors of North America, Europe, and Asia.

There are many different types of snow. In English, there is only one word for it, but in some languages there are several. Some snow is powdery and dry, and some is heavy and wet. Some crystals are as fine as flour, while others can stick together to form true flakes measuring an inch or more across. The finer crystals tend to develop when it is very cold; the large, wet flakes form at near-freezing temperatures. For this reason, an early-winter or late-winter snow is likely to be wet. Powdery snow falls most often in the "hard winter" months.

MOISTURE CONTENT OF SNOW

The moisture content of new-fallen snow is expressed in terms of the amount of water after melting. For dry, fluffy, powdery snow, the water-equivalent ratio is between 10:1 and 15:1. This means that 10 to 15 cm (4 to 6 in) of snow will melt down to 1 cm (less than 0.5 in) of water. Wet snow has a ratio as small as 5:1. The wet snow is not only stickier, but it is more dense. This is why a given depth of wet snow is heavier than the same depth of dry snow.

If conditions are just right, an early-season or late-season snow storm can cause significant damage to trees and utility lines. The effects are similar to those of an ice storm. A particularly severe storm of this type took place in southern Minnesota in March of 1966. A heavy rain changed to wet snow late one evening. Meteorologists warned that the total snow accumulation might exceed 25 cm (10 in). The flakes piled deeper by the minute, accumulating at the rate of more than 2.5 cm (1 in) per hour. By morning, 28 cm (about 11 in) of snow had fallen. It clung to everything. Tree damage was extensive. The situation was worsened by gale winds that followed the snowfall. Utility poles snapped as the wind blew against wires coated with up to 5 cm (2 in) of sticky, dense snow. In some locations, series of three, four, or five poles in a row were left standing upside down next to their broken-off stumps, while the utility lines themselves were unbroken!

Saturated snow can do more harm than merely damaging trees and downing power lines. The winter of 1977–1978 produced snows so heavy in New England that several roofs collapsed. The most sensational of all was the ruin of the Civic Center in downtown Hartford, Connecticut. That casualty was caused by the weight of already wet snow further saturated by rain.

Heavy, wet snow is harder to remove from roads and driveways than fluffy, dry snow (except when blowing and drifting occurs with dry snow, constantly foiling the efforts of snow removal crews). Wet snow chokes snowblowing machines, making it necessary to use shovels to remove it from sidewalks and driveways. This can cause injuries and even deaths among middle-aged and elderly people who attempt to shovel the snow when they are "out of shape" (have not been getting enough exercise as a routine).

PROBLEM 7-1

Why are winters in International Falls, Minnesota so much colder than winters in Seattle, Washington, even though the two towns are at approximately the same latitude?

SOLUTION 7-1

The Pacific Ocean moderates the temperatures in Seattle. In Seattle, the prevailing winds come from the west, having passed over thousands of kilometers of open sea before making landfall. In International Falls, the prevailing westerlies have passed over thousands of kilometers of land mass, where the air gets a chance to cool down much more than is the case over the ocean.

PROBLEM 7-2

Some autumn seasons pass gradually into winter, while other autumn seasons yield suddenly to winter. We've heard about all sorts of factors that might affect this, such as ocean currents, sunspot cycles, and volcanic activity. Is there a single "prime mover" that drives cycles of winter weather?

SOLUTION 7-2

No one knows. It is tempting to suppose that climate variations may occur in cycles of several years or decades. But occasional "freak seasons" fail to fit into any predicted pattern or cycle.

Anatomy and Effects

A winter storm is a low-pressure system that covers a large geographic area and contains weather fronts. The circulation is counterclockwise in the northern hemisphere, and clockwise in the southern hemisphere.

STORM STRUCTURE

In a typical winter storm, a cold front extends toward the equator from the center of the system, and a warm front runs in a generally east–west direction on the eastern side of the system. Rain usually falls between the warm and cold fronts, except during the coldest part of the season, when it can be rain, ice pellets, freezing rain, or snow. Behind the cold front, high winds often occur, and there is light snow or no precipitation. It is here that *blizzard* conditions occasionally arise. On the polar side of the system, snow falls. In a large storm, this is where the heaviest accumulations take place.

Figure 7-3 shows the typical patterns for northern-hemisphere and southern-hemisphere winter storms. In these examples, the storm tracks are generally from west to east, although the paths of actual storms vary between compass headings of approximately 30° (toward the north-northeast) and 150° (toward the south-southeast). The frontal patterns shown in Fig. 7-3 are typical, but there can be considerable variation. For example, the cold front may catch up to the warm front near the center of the system, forming an occluded front. A second, stronger cold front may develop behind the original one, particularly if the storm is large.

STORM FACTORS

Three primary factors determine the type of weather that accompanies a winter storm at a given place:

- The temperature
- The storm orientation with respect to the locale
- The direction of storm movement

The overall temperature dictates the relative geographical sizes of the snow area and rain areas. Early-season and late-season storms usually have rain everywhere, except far to the poleward side of the center. Mid-season storms have

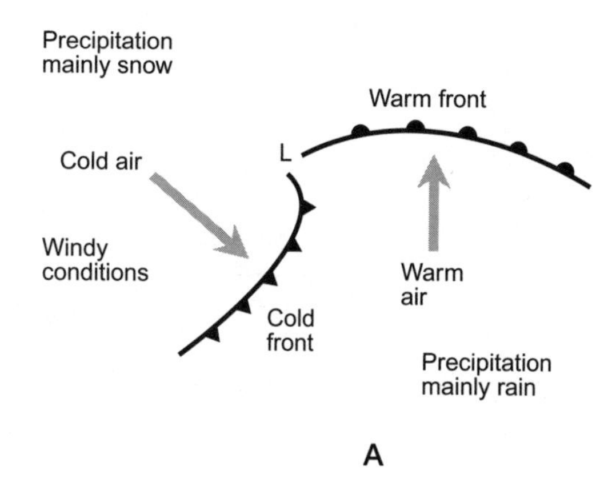

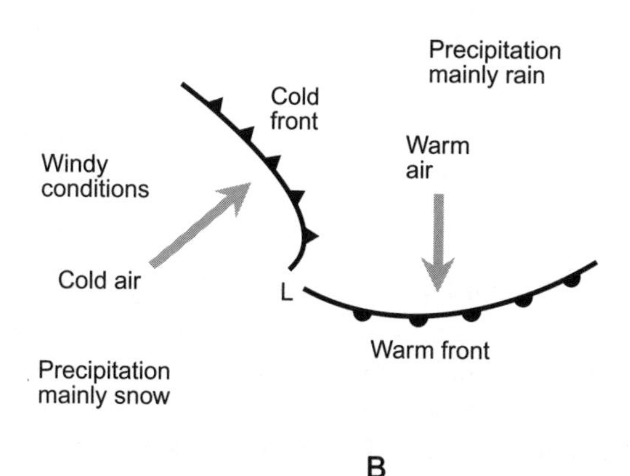

Fig. 7-3. Typical winter storm systems over land in the northern hemisphere (A) and in the southern hemisphere (B). In both of these examples, the overall storm movement is generally west to east (left to right in these drawings).

smaller rain areas and larger snow areas, and in some of these storms the precipitation is all snow. Latitude affects the temperature of a storm system, as does the proximity of an ocean. Storms over land are nearly always colder than storms over the oceans. If an ocean storm comes inland, it cools down; if a continental storm moves out over the ocean, it warms up.

The center of a storm moving from west to east (compass heading 90°) may pass to the north of a specific location on the surface, to the south of it, or right over it. These three situations are shown in Fig. 7-4 for a northern-hemisphere

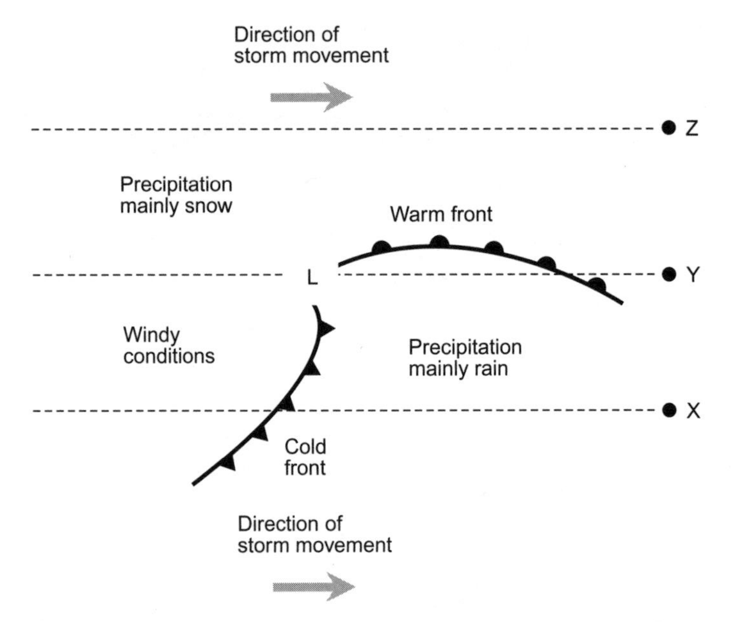

Fig. 7-4. As a winter storm passes three different towns (X, Y, and Z), the sequence of events in each town is determined by the part of the storm that strikes.

storm. Town X lies to the south of the storm track. The first precipitation will be rain, accompanied by winds generally out of the south. As the cold front nears and then passes, the rain will change to snow and taper off. There is a good chance of high winds behind the cold front. Town Y lies directly in the path of the storm. The weather there will be similar to the sequence of events at X, except that we can expect heavier rain or snow in advance of the passage of the center, because of the warm front. Town Z will get heavy snow, along with winds that back from southeast to east to northeast.

The third factor that determines how a storm will behave is the direction in which it moves. The storm of Fig. 7-4 would produce different weather at any of the three towns shown if the movement was toward the north-northeast (compass heading 30°), for example, instead of toward the east.

RAIN INTO SNOW

During a winter storm, the precipitation sometimes changes from rain to snow as the cold front passes and the temperature drops. The wind characteristically veers from south or southeast to west or northwest as the changeover takes place.

The rain-into-snow type of winter storm is common in the Midwest, because the jet stream normally flows eastward or southeastward over these regions. Cold air prevails north of the jet stream, and warm air prevails south of it. Storms tend to track along the jet stream, usually centered slightly to its north.

Figure 7-5 shows two typical midwinter jet stream paths over the United States. At A, a ridge lies offshore in the Pacific, and a massive trough dominates the center of the continent. This brings moderate weather to the west, and severe cold in the Midwest. The Atlantic coastal area escapes the grip of the cold. At B, a strong ridge prevails over the western United States with a deep trough over the eastern part of the country. This results in warm, dry weather in California and Arizona, while all of the eastern states have unusually cold conditions.

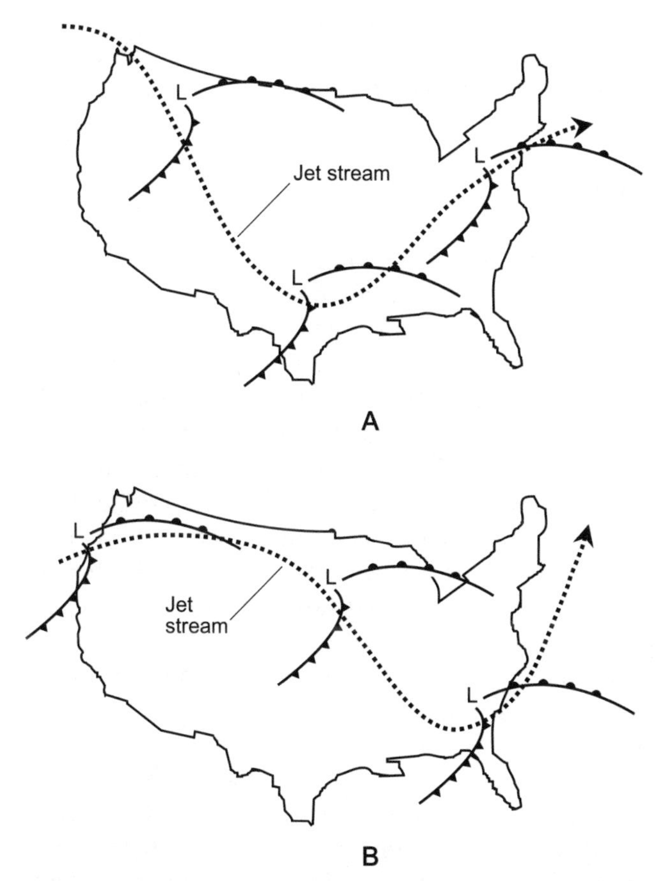

Fig. 7-5. Two hypothetical wintertime jet stream patterns over the United States. At A, a central trough; at B, a ridge in the western United States with a trough in the east.

Consider the situation in a Midwestern town, located under the jet stream or just to the south of it as a winter storm passes. The first sign of the approaching low-pressure system is a gradual increase in high cloudiness. The clouds first appear as cirrus or cirrostratus, and thicken until the sky is dark and gray. The temperature remains constant or rises slightly as the warm front passes. The wind blows from the south or southeast. Rain falls behind the front. As the cold front approaches, the wind veers, and the temperature drops. When the front arrives, the temperature drops below freezing, and the rain changes to snow. Wet roads acquire a layer of thin or patchy ice as the water on them freezes, and driving conditions become treacherous. When the cold front moves through the area, the wind shifts to the west or northwest and increases in speed. Visibility deteriorates as the snow begins blowing and drifting. If the storm system is intense, blizzard conditions, with dangerous cold and high winds, occur. If the jet stream pattern is similar to that shown in Fig. 7-5A, snow can fall as far south as northern Mexico. If the flow is more like that of Fig. 7-5B, the Great Plains states will be hardest hit, and snow will blanket much of the southeast.

SNOW INTO RAIN

Along the eastern seaboard during the winter, the jet stream flow is often, if not usually, from the south-southwest or southwest toward the north-northeast or northeast (compass heading 30° to 45°). This, along with the proximity of the Atlantic Ocean, brings a different sequence of events to this part of the country when a winter storm passes, as compared with the situation in the Midwest.

Suppose you are located in Connecticut as a low-pressure system follows the jet stream path shown in Fig. 7-5A. The first signs of the storm are high clouds, such as cirrostratus, along with an increasing easterly or northeasterly wind. The clouds gradually become thicker and lower, and snow flurries begin. As the storm center approaches, the winds increase, and the snow falls heavily. A strong storm can produce gale-force winds and high tides along the coast. When the center of the low-pressure system passes to the west and northwest of Connecticut in the hypothetical situation of Fig. 7-5A, the wind veers to a southerly quarter, and a warm front arrives. The temperature rises, and the precipitation changes to mixed snow and rain, and finally it is all rain. Only in an extremely cold system does the precipitation remain as snow following the warm front.

New England is famous for its ice storms, and the snow-into-rain scenario of Fig. 7-5A presents ideal conditions for such an event. As the warm front passes, all of the trees, utility lines, cars, and rooftops are still cold. If the temperature rises from well below freezing to just above freezing, the raindrops will turn to ice as soon as they land. Residents of New England have learned to carry cans of lock de-icer in their coat pockets during the winter months. Sometimes rain makes its way into the door lock mechanisms of a car and then freezes. Lock de-icing compounds are available at most good auto supply stores, and they can be used to thaw out such ice in most situations—as long as the can of de-icer is not locked in the car!

BIG SNOW MAKERS

If the center of a low-pressure system passes off to the southeast and east of New England, a different thing happens. A situation of this kind often develops when the jet stream is situated in a manner similar to that shown in Fig. 7-5B. The Atlantic waters provide a rich supply of moisture for the circulation, because much of the storm system lies offshore. Heavy snow develops. Inland, the wind backs from northeast to northwest.

In January 1978, an Atlantic Coast storm passed Cape Cod and then veered out to sea. Most of New England was buried in 30 cm (1 ft) or more of snow. Hartford got 36 cm (14 in). A few days later, another system, even more severe, followed a similar path. Up to 1 m (39 in) of snow fell in parts of New England. Boston received that much snow, and was also battered by winds that gusted to 80 kt (about 90 mi/h). A few whimsical people put on downhill skis and raced along level streets, propelled by the wind. The nearby shores were inundated by a storm surge, as if the system were a hurricane. In fact, on the satellite photos, the storm (dubbed "Larry") bore some resemblance to a hurricane, even showing a localized eye.

As bad as it was, storm Larry could not compare with the "nor'easter" of March 12, 1888. From Washington, D.C. to the tip of Maine, the region was immobilized by snow accumulations of up to 1.27 m (50 in). Some drifts were as deep as 10 m (33 ft).

Large bodies of water always increase the snow-dumping capability of a winter storm. Monsters such as the "nor'easter" of 1888 and Larry of 1978 demonstrate this well. The Atlantic Ocean, however, is not the only attractive feeding ground for snowstorms.

LAKE-EFFECT SNOW

As the glaciers retreated during the waning centuries of the most recent ice age, the present-day landscape was carved into the central and northern parts of North America. The glaciers left five large freshwater lakes—the *Great Lakes*—in the heart of our continent: Superior, Michigan, Huron, Erie, and Ontario. They are famous for the snowstorms they spawn. Heavy snows regularly fall in northern Wisconsin, northern Illinois, Indiana, Ohio, Pennsylvania, and New York state, as well as a large part of Michigan. The most dramatic accumulations occur along the southern and eastern shores of the lakes. This is called *lake-effect snow*.

Lake-effect snow events commonly take place following the cold front of a low-pressure system. A strong wind, blowing from the northwest toward the southeast, howls across the large expanse of open water, picking up moisture. The water precipitates as snow over the land areas on the leeward sides of the lakes. Great Lakes snowstorms can produce accumulations well in excess of 1 m (39 in) within a couple of days. The problem is often compounded by bitter cold temperatures and by high winds. This creates dangerous blizzard conditions.

A series of storms spawned by Lake Erie in 1977 gave the city of Buffalo, New York, a massive blanketing of snow. The dry, powdery snow drifted completely over cars and trucks, whipped along by winds that gusted to about 70 kt (80 mi/h). The temperature fell far below freezing. Visibility was zero.

In recent decades, the consequences of lake-effect snowstorms have been increasing because of the activities of humans. There are two reasons for this. First, the shores of the Great Lakes, especially on the United States side (which is usually leeward, where most lake-effect snow falls) are peppered with cities and industrial complexes, so there are millions of people in the affected region. Second, the Great Lakes have been getting warmer because of industrial heat pollution. This warmth has made the lakes better "food" for the snow-hungry winds that sweep down from Canada during the winter. Ironically, some of the cities most responsible for the heat pollution are the hardest and most frequently hit.

MOUNTAIN SNOW

The Atlantic and the Great Lakes can produce giant snowstorms because of the interaction among wind, water, and land, but winter has still another ingredient for snow making: mountains. When moisture-laden air sweeps up the side of a mountain range, the resulting snow accumulations can be hard to believe.

The Berkshires in New England, and parts of the Appalachians, are well known for their deep snows. The ski resorts of Colorado, Utah, and other Rocky Mountain states are also famous for their snow. But to see nature's snow-making machine at its finest, we must go to the Pacific Coast. A good example is Mount Rainier, near Seattle. Another example is the Lake Tahoe region. In January, 2004, this resort area received a series of storms resulting in snow accumulations of up to 6 m (approximately 20 ft). The same thing occurred again in 2005, and the effect spilled past the mountains into the Carson Valley and Reno, Nevada, where the snow depth reached more than 2 m (6 ft).

Along the west coasts of continents and islands at latitudes between approximately 35°and 60°, the prevailing winds blow in from the ocean, resulting in copious amounts of precipitation. The three largest areas of this type in the world are coastal Europe, a large portion of Chile in South America, and much of the west coast of North America. The temperate west coasts of North and South America are mountainous, and the mountains draw the moisture out of the air by a process known as *orographic lifting*. As the moisture-laden air rises, it cools, and the temperature falls below the dewpoint. Some of the moisture in the air precipitates out. As a result, the Pacific Northwest gets abundant rainfall near shore, and abundant snowfall on the windward sides of the mountains (Fig. 7-6).

At the latitude of Mount Rainier, the temperature near the summit is almost always cool enough to result in snow. Snow falls all year long. The largest amounts occur in the colder months, when Pacific-spawned storms are the most intense. In a single year, the snowfall on the slopes of Mount Rainier can exceed 25 m (about 80 ft), and individual storms routinely drop 2 m (6 ft). Similar conditions exist up and down the coast from central California to Alaska. The same is true along a sizable stretch of the South American Andes. The coastal regions of Europe receive less snowfall because the region is less mountainous, although plenty of heavy snowstorms take place in the Alps and other inland mountain ranges.

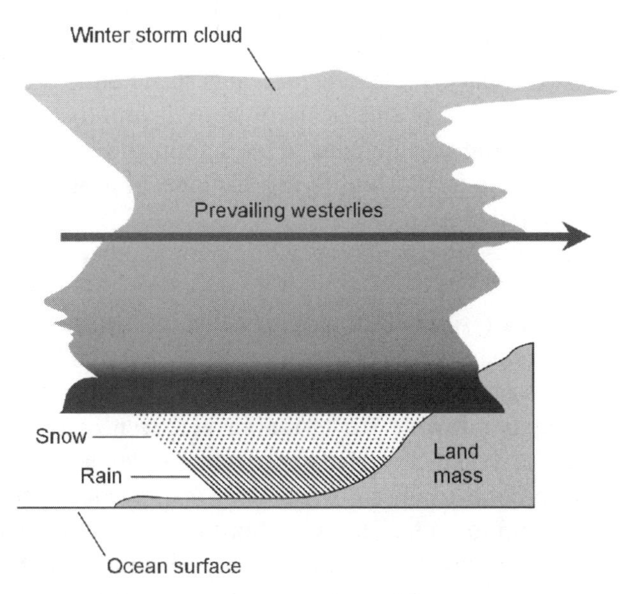

Fig. 7-6. As moist air blows in from the ocean during the winter and encounters mountainous land, rain falls at the lower elevations, and snow falls in the mountains.

AVALANCHES

We sit in our mountain cottage on a sunny morning. The ground glistens with new-fallen snow. We have a hot breakfast and get ready for the work of shoveling our way outside. Snow blocks the doorway, and we'll have to climb out of a window to clear it. The roof, too, needs to be shoveled off before the snow begins to freeze around the eaves.

We hear a sound like thunder in the distance. It's too cold for a thunderstorm. Perhaps it is a jet airliner. We walk over to a window and gaze up at the slopes. Then we see what is making the noise: A whole section of the mountain is in motion. A white cloud of snow is racing toward the valley below. It is an avalanche. We have all heard stories about monstrous avalanches. Every ski season, a few people get hurt or killed by this phenomenon. Avalanches take place in the mountains where snow accumulates to a significant depth and is subjected to certain forces.

Avalanches seldom occur on smaller hills, and never on flat ground. The snow pack on a mountain in the middle of the winter may appear to be stable, but it has been built up as the result of several individual storms. Different storms produce slightly different types of snow, and the snow also changes with time under the pressure of its own weight. The result is that the snow pack on a mountain slope accumulates in layers during the course of a winter. Thus, it is not uniformly stable. This wouldn't cause a serious problem all by itself on flat land, but on a sloping surface the situation is different. Gravity pulls not only straight down on the snow, but sideways (Fig. 7-7). If the force is great enough, and a certain snow boundary or *cleavage plane* is sufficiently weak, a small disturbance can cause a section of one or more snow layers to break off and start sliding. The initial rupture can be caused by a falling tree, a passing skier or snowmobile, the vibration from a highflying airplane, or a small animal hopping across the snow-covered slope.

After a fracture forms, it can spread with amazing speed, sometimes by as much as 250 kt (about 290 mi/h). The fracture may grow to be more than 100 m (330 ft) in length. The top snow layer alone, or several layers, might be involved in the avalanche. The disturbance of the initial snow movement can cause other large pieces of snow to begin avalanching.

As the snow catapults down a slope, the air is driven before it, producing a wind gust ahead of the avalanche. The wind itself can level trees and hurl loose objects through the air. In extreme cases, the wind acquires the speed and wrenching force of an F2 or F3 tornado. The wind blast covers a much larger area than the avalanche itself, and presents a significant danger to life and property.

In ski resort areas, people try to prevent avalanches by releasing unstable snow packs before they get large enough to be dangerous. This is done by using

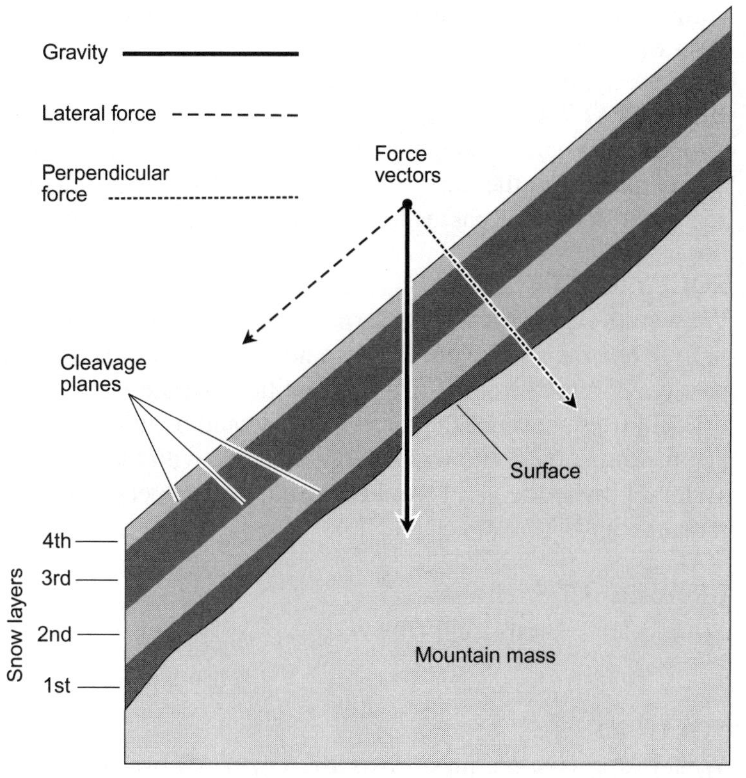

Fig. 7-7. Gravity produces lateral tension along the cleavage planes between layers of snow on a slope. A small disturbance can thus produce a large avalanche.

explosives on slopes known to be avalanche prone. A few small, deliberately instigated avalanches are better than a single "maxi avalanche" that strikes a populated place.

PROBLEM 7-3

If all other factors are equal, how does the slope of a mountain affect the likelihood of avalanches?

SOLUTION 7-3

As you would expect, the steepness of the slope has a direct effect. Avalanches do not occur on horizontal surfaces, because the lateral component of gravitational force is zero. As the grade becomes steeper, the lateral force component increases, placing more and more stress on the boundary planes where different layers of snow meet. On a near-

vertical surface, an avalanche can occur even if there is only a single layer of snow.

PROBLEM 7-4

Suppose the jet stream has an unusual bend, and a winter storm approaches from the north-northwest, moving toward the south-southeast. Its center passes to our east. What can we expect?

SOLUTION 7-4

We would expect relatively warm conditions and a westerly wind initially. Then the wind would veer to northwest or north, attended by the passage of a cold front. Blizzard conditions would likely occur behind the cold front, because the temperature would be low enough for snow, and the direction of the wind would coincide with the movement of the system. Finally the wind would slacken, but temperatures would likely remain frigid.

PROBLEM 7-5

What is an "Alberta clipper"?

SOLUTION 7-5

Winter storms sometimes cross the upper Midwestern states (North Dakota, South Dakota, Minnesota, Iowa, and Wisconsin in particular) from northwest to southeast. These storms are called "Alberta Clippers" because they move fast, produce high winds, and come from the general direction of the Canadian province of Alberta.

Blizzards

During the winter months, storms spin off from the Aleutian and Icelandic low-pressure systems. The two permanent lows reach their greatest intensity in January and February. The prevailing winds carry the spinoffs away, and they eventually make landfall. In North America, the centers of the winter storm systems come ashore between southern California and southern British Columbia. Although winter storms produce relatively little snow along the immediate Pacific Coast, high winds and heavy rains are common. Wet weather can penetrate far inland, drenching the deserts and steppes.

JET STREAM SOUTH

When the jet stream is displaced southward from its average position over the western coast of the United States and a strong storm comes ashore from the Pacific, the deserts of California and Arizona temporarily become monsoon zones. Unaccustomed to the sudden climatic change, rivers overflow their banks. Strong onshore winds can cause abnormally high tides with resulting damage to property. It seems that, almost every year, we read or hear stories about luxurious seaside California homes sliding down mud banks into frenzied water.

As the storm system moves eastward, cold air from the continental Arctic begins to feed the circulation. The rain may change to snow, covering the plains of Texas and Kansas with wind-driven snow before continuing on to the southeast and east. The upper Midwest may get heavy snow, followed by bitter cold. Ultimately, the storm follows the jet stream up the east coast, making things sloppy for residents of New England, and then moving out into the waters of the Atlantic.

A similar sequence of events takes place in Europe. If the jet stream is displaced to the south of normal, the normally mild climates of Spain and southern France turn foul. The waters off the South of France are well known for their powerful *mistral* storms, caused by north winds following cold fronts blowing off the continent. A system from Europe typically moves into Asia Minor. Several storms of this kind, following each other in rapid succession, caused protracted winter storm conditions in Yugoslavia during the 1984 Winter Olympics.

JET STREAM NORTH

If the jet stream is located at, or to the north of, its average position over the west coast of North America, the California and Arizona deserts are spared the effects of Pacific storms. Their residents enjoy the mild, sunny winter weather for which the region is famous. As a storm system makes landfall under these conditions, rain and gales pound the coasts of Washington and Oregon; then the storm moves into the Rockies, across the central plains, the Ohio Valley, and New England.

When the jet stream moves northward over Europe, winter storms move inland in the vicinity of the British Isles, northern France, and Germany, and continue on into Russia. These systems commonly turn into blizzards. Russian winter storms are as fierce as their American Midwestern counterparts. Napoleon in the early 19th century, and Hitler in the middle of the 20th century, learned at great cost that the Russian winter could be an insurmountable adversary. They tried to conquer Russia and failed, in part because of the severity of the winter.

Blizzards and cold temperatures strike early in the plains of eastern Europe because of their distance from the moderating oceans. The climate of the region is similar to that of the Dakotas and Minnesota. In 1941, the year of Hitler's ill-fated military campaign in Russia, the winter arrived early, with unusual severity. Heavy rains came first; fields and roads turned to mud. Then the temperature dropped below freezing, and the mud hardened like concrete. Finally the snowstorms began, marching across the continent one after another. The first snows came on October 6 and continued for a week. By December, temperatures had reached −40°C (−40°F). The heavy snow and bitter cold were accompanied by gale-force winds that lasted for periods of days without interruption. Some historians believe that had the winter of 1941–42 been mild over the plains of western Russia, the Germans would have reached and taken over Moscow.

WINDCHILL FACTOR

The cooling power of the air is increased by wind. Exposed flesh, as well as other warm or hot objects, are affected more severely as the wind speed increases at a given temperature. The effective temperature, taking the wind into account, is called the *windchill factor*. At a given actual temperature, the windchill factor is the air temperature that would cause the same amount of cooling to exposed flesh as would occur if the same person were walking at a normal pace in calm air. It is assumed that there is no sunlight to heat the skin, and that the body temperature is normal.

The windchill factor is a subjective expression of equivalent temperature, because people do not walk around naked in raging blizzards. Moreover, the windchill function is not the same for machines such as automobile engines and electronic equipment, which have operating temperatures different from that of human flesh. The windchill temperature is, however, a valuable guide to the relative danger of combined cold and wind. It is especially important to know the likelihood of frostbite to hands and face.

A significant danger of frostbite exists if the windchill factor is below −30°C (−22°F). The cold can do some harm, however, at any temperature below freezing. An unprotected person can die of exposure at temperatures significantly above freezing if the wind is strong. In extreme conditions, the risk is obvious, and residents of the blizzard-swept Northlands of the United States have coined a little rule known as the "30-30-30 law:" at −30°F (−34°C) in a 30-kt (35 mi/h) wind, bare skin freezes in 30 s. On the Russian front during the hard winter of 1941–42, conditions were like that, or worse. Similar brutal conditions have caused civilian deaths in the United States. In parts of North Dakota, Minnesota, and Wisconsin, for example, windchill factors sometimes drop as low as approximately –70°C (–94°F).

It does not take very much wind to cause a windchill factor that is far lower than the actual temperature. When the wind reaches gale force, the windchill factor for a given temperature "bottoms out." Then higher winds will not, for a given temperature, reduce the windchill factor any further.

WHAT QUALIFIES AS A BLIZZARD?

Meteorologists define blizzards according to the temperature and the wind speed. Blizzards are characterized by blowing and drifting snow, limited visibility, and cold temperatures. When the temperature drops below –7°C (+25°F), accompanied by a wind of 30 kt (35 mi/h) or more, along with blowing snow, *blizzard conditions* exist. If the temperature falls to less than –12°C (+10°F) with winds of 39 kt (45 mi/h) or more, *severe blizzard conditions* exist.

Although the meteorologist does not rigorously define more severe categories of blizzards, we might do well to consider that conditions can get a lot worse than those just described. Winds in some blizzards have been known to gust to over 100 kt (115 mi/h), and temperatures can fall below –40°C (–40°F) in parts of the world normally considered temperate. When a winter storm attains such magnitude, travel is nigh impossible. People caught unprotected find themselves in mortal danger.

Blizzard conditions do not necessarily come in conjunction with heavy snow. A snowfall of 50 cm (20 in) produces more inconvenience than an accumulation of 5 cm (2 in), and a heavier snow naturally increases the difficulty of travel. It is not the snow, however, that presents the greatest risk to life. It is the extreme cold, exacerbated by high winds. A blizzard with light snow can be more dangerous than one with heavy snow, because people are more likely to attempt travel when the snow accumulation is not great. They forget that the snow can drift. A 5-cm (2-in) snowfall, whipped along by gale-force winds for a few hours, can pile up into drifts that are invisible in the storm, but deep enough to stop a truck.

INGREDIENTS OF A BLIZZARD

A cyclonic bend in the jet stream—a turn from right to left in the northern hemisphere—can cause winter storm systems to intensify to the point where blizzard conditions develop. The counterclockwise flow of air enhances the circulation of a low-pressure system, increasing the speed of the winds. Also, a cyclonic bend is likely to be accompanied by *atmospheric divergence,* which causes the barometric pressure to drop more than it otherwise would. Cyclonic bends in the jet stream are common over the United States and eastern Europe during the colder months.

A blizzard can also develop from a low pressure system that is unusually strong from the very beginning. This type of storm produces high winds near and behind the cold front. Thus the lowest temperatures and the highest winds tend to occur in the same part of the storm. Most blizzards reach their maximum intensity, in terms of wind speed, once they have arrived at the point in the jet stream nearest the equator. The storms can remain severe, however, for the remainder of their eastward or northeastward trek across the continent. In North America, as a blizzard nears the Atlantic coast, the snow accumulation increases, but the temperature moderates.

Blizzard conditions sometimes occur following a winter storm that we would not consider a true blizzard. An exceptionally strong high-pressure system can produce bitterly cold temperatures and winds that gust to gale force. The highest winds blow on the northern side of the center, where the movement of the system complements the wind circulation. The result, if snow has just fallen, is similar to a blizzard even though there is no precipitation. The sky may be clear, but visibility is severely restricted within a few meters of ground level. Such *ground blizzard* conditions can be dangerous because, although the weather doesn't look too bad, the drifting snow and low windchill factor produce conditions as adverse as those in a more fierce-looking storm.

BLIZZARD WATCHES AND WARNINGS

Meteorologists issue *winter storm watches* and *winter storm warnings* in much the same way as they do for tropical storms, hurricanes, severe thunderstorms, and tornadoes. If a hurricane, severe thunderstorm, or tornado watch is posted, the forecasters do not mean to imply that any specific location will be directly hit. The situation is different in the case of a winter storm. Winter storms are much larger than any other type of storm in terms of the sheer geographic areas they strike. A winter storm watch is frequently followed by a warning, and then by winter storm conditions.

A *blizzard watch* can be issued for either of two different reasons: when a winter storm is expected to mature into a blizzard and then strike a given place, or when the danger is high that a blizzard will strike a given place within a day

or two. If a storm has matured into a blizzard and is expected to strike within 24 hours, a *blizzard warning* is posted. A *severe blizzard warning* is issued if winds of more than 39 kt (45 mi/h) are expected in conjunction with temperatures lower than −12°C (+10°F).

You can tell, to some extent, how severe a blizzard will be by checking the barometric pressure as the center of the storm passes nearby. If the weather maps in your local newspaper contain information about the barometric pressure, you can check the figures at the storm center. An excellent source of weather information is available on the Web at The Weather Channel (www.weather.com). If a winter storm has a central pressure of 1000 mb (29.5 in of mercury) or less, you can expect a blizzard. If the pressure is lower than 992 mb (29.3 in), you should be prepared for a severe blizzard. Occasionally, blizzards develop central pressure below 980 mb (28.9 in). If you see this kind of barometer reading, you can expect extreme blizzard conditions, with heavy snow and sustained gale-force winds.

If the temperature is cold enough, the snow in a severe blizzard can become so fine that it finds its way inside gloves, mittens, boots, and overcoats as if it were dust. These "Dakota blizzards" are legendary for their lethal combination of bitter cold, highway blockages, power outages, and zero visibility.

BLOWING AND DRIFTING SNOW

When windblown snow reduces visibility and becomes packed into hard drifts, driving conditions become poor. The wind sometimes sculpts the packed snow into strange formations that hang from the eaves of buildings. In some blizzards, snow has been known to drift so high that whole houses are buried. Of course, the greater the amount of snow that falls, the deeper the drifts will be, all other factors being equal. But in a prolonged blizzard with gale-force winds, it does not take much snow to produce heavy drifting.

When all of the ingredients of a blizzard—wind, cold, and snow—combine for several days, the activities of civilization come to a halt. Buffalo, New York, experienced a situation like this in the winter of 1976–77. Snow drifting can be controlled to some extent by means of barriers. Drift control is especially useful to prevent roads from becoming impassable. Temporary fences, especially designed for drift control, are used in the American Midwest to help keep roads clear. The blowing snowflakes slow down as they pass through the vertical wooden slats of the fence (Fig. 7-8A). This creates a drift on the lee side of the fence. Because the worst blowing and drifting in Midwestern blizzards are usually caused by west or northwest winds, snow fences are erected on the north sides of roads running east and west, and on the west sides of roads running north and south (Fig. 7-8B). The fences are positioned far enough from the roads so that the artificial drifts are formed before the snow gets to the road surfaces. When spring arrives, the fences can be rolled up and put in storage. Permanent barriers, consisting of rows of evergreen trees on either side of the road, are also frequently seen in regions where blizzards are common.

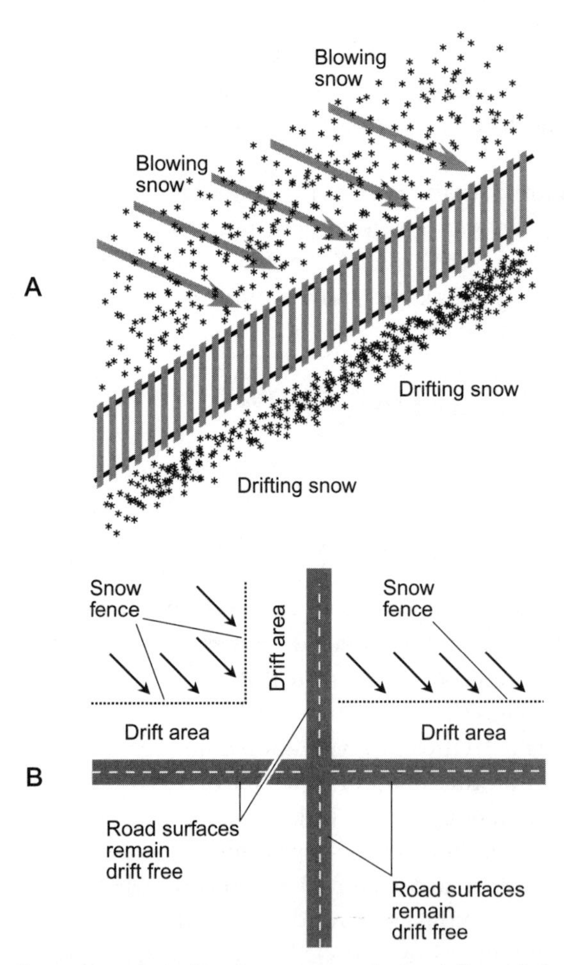

Fig. 7-8. A snow fence slows down blowing snow, producing drifts and clearing an area downwind (A). In blizzard-vulnerable regions of the American Midwest, snow fences are placed along the north and west sides of roads (B). Arrows represent the wind.

AFTERMATH OF A BLIZZARD

Following a heavy blizzard, the cleanup task begins. Mailboxes and fire hydrants are dug out first, and then the snow plows get busy. Before the proliferation of motor vehicles in civilized countries, our modern, frenzied snow-removal programs would have been considered ridiculous. Decades ago, sled-type runners replaced wheels on vehicles until the snow melted in the spring, but today, millions of dollars are spent every year to clear snow off of streets and highways. An incredible variety of plows, blowers, and shoveling devices keeps our modern transportation system moving.

Sometimes, despite all of our snow-removal machinery, a blizzard defeats us. We clear a spot, and a few minutes later it is drifted in again. Nature forces us to take a day or two off and slow down. It's difficult for some adults to tolerate this, but children don't seem to have a problem with it, especially if school is canceled.

STAYING SAFE

People can die in snowstorms and blizzards. Some insist on travel when roads are impassable, some attempt to shovel a sidewalk or driveway in spite of the fact that they never get any exercise, and once in a while someone actually gets lost between a house and a barn or garage.

When people are injured or die in winter storms, there are usually complicating factors. Most people who lose fingers, toes, or limbs in blizzards suffer from *frostbite.* Death is nearly always the result of *hypothermia.* Frostbite occurs when skin actually freezes, or when the circulation is restricted because of excessive cold. Hypothermia is an abnormally, and pathologically, low body temperature. If a blizzard is in progress, it is unwise to venture out in it, but in some cases people believe they have no choice. Proper clothing then becomes crucial for protection against frostbite and hypothermia.

The chief warning sign of frostbite is numbness or whiteness of the skin, especially in extremities such as fingers, toes, and ears. For protection against frostbite in the fingers, mittens are better than gloves. Feet can be kept warm with two or three pairs of socks, soft shoes, and rubber boots ("galoshes") worn outside the shoes. For ear protection, a thick fur cap works well. A parka with a hood is better. It is important that hands, feet, and head be kept dry.

Hypothermia causes confusion and drowsiness. The risk of hypothermia can be minimized by dressing in layers, as well as by taking all of the precautions for guarding against frostbite. It also helps to make sure you have had enough to eat. The body's demand for calories is increased in cold weather. All high-calorie foods

(except alcohol), consumed in moderate quantities, help to maintain body temperature. For most people, choosing an appealing high-calorie food is not a problem.

Alcohol increases the danger of hypothermia, largely because it impairs judgment. It is difficult enough to get around in a blizzard without getting drunk. Alcohol also causes the small blood vessels in the skin to dilate, which increases the rate at which body heat is lost.

BLIZZARDS AND MOTOR VEHICLES DON'T MIX!

Most of the deaths in blizzards occur in conjunction with motor vehicles. Poor weather conditions increase the number of single-car and multiple-car accidents per unit volume of traffic. Also, there is a risk of getting stranded.

The number one rule for driving in a blizzard is this: Don't, unless there is a genuine emergency! Ideally, all travel should be suspended during a blizzard and resumed only when the storm is over and the snow has been cleared from the roads. *Snowmobiles,* also called *snow machines,* are made especially for travel in severe winter weather, but few people own them, and they are therefore forced to use more conventional motor vehicles.

For traveling in a blizzard or snowstorm, larger cars are better than smaller ones. Pickup trucks are better still. Vehicles with high ground clearance are better than cars with low ground clearance. If it becomes necessary to drive in a blizzard, certain supplies will be of help in case you get stuck:

- A working cell phone with a fully charged battery
- An amateur or citizen's band radio transceiver
- A shovel to clear away snow
- One or more warm blankets
- Several bags of sand, placed in the trunk
- Tire chains
- A flashlight or lantern with spare bulb and spare battery
- A supply of storable, high-calorie food
- A jug of water to alleviate thirst
- A large thermos bottle of hot soup or coffee

Even if the roads are free from ice, windblown or tire-packed snow can be slippery. Here are some common-sense safety rules for driving in blizzards:

- Do not consume alcohol
- Do not consume narcotic drugs
- Keep the gas tank at least 3/4 full at all times

- Use headlights (low beam) even during the day
- Drive at a slow speed (but not too slow)
- Follow other vehicles at a greater distance than usual
- Accelerate and brake gradually
- If a skid occurs, release brakes and steer in the direction of the skid
- Use a low gear when driving up or down steep hills

Every year, people get stranded in blizzards with their cars. The experience can be frightening. We are used to warm, comfortable cars and homes. Many people spend whole days indoors during the winter, except for the minute or two that it takes to run from the parking lot to the office in the morning, and from the office to the parking lot in the evening. It is possible to go from one place to another in dangerous cold and not spend more than a few seconds outside. It's enough to foster overconfidence, but that secure feeling vanishes quickly for the passengers of an automobile that gets stuck in a blizzard in some isolated place. The following safety rules can be of value to people who are marooned in a severe snowstorm or blizzard:

- Passengers should stay with the vehicle unless a source of assistance is within sight
- Passengers should stay inside the vehicle as much as possible
- Attempts can be made to free the vehicle, but overexertion should be avoided
- The engine should be run for only about 10 minutes per hour
- The exhaust outlet should be kept clear to prevent engine stalling
- A downwind window should be kept open a few centimeters for ventilation
- Snow should not be eaten because it reduces body temperature
- The emergency flashers (but not the headlights) should be on
- Inside lights should be shut off
- Alcohol and drugs should be absolutely avoided, except necessary prescriptions
- After the rescue, everyone should get medical attention

PROBLEM 7-6

Stories have been told about some famous blizzards in which the visibility was reduced to the point that "you couldn't see your hand in front of your face in the middle of the day." Isn't that an exaggeration? How is that possible?

SOLUTION 7-6

There are at least two reasons for this. First, with temperatures far below zero and winds that sometimes gust to speeds in excess of 70 kt

(80 mi/h), it is difficult or impossible to keep your eyes open unless you look directly away from the wind. Second, at extremely low temperatures, blowing snow becomes as fine as dust. For a given rate of snowfall in centimeters per hour (cm/h), and for a given wind speed, visibility is proportional to the size of the snowflakes. A frigid, violent ground blizzard thus resembles a "frozen dust storm" and can reduce actual visibility to a few meters.

PROBLEM 7-7
What is a whiteout?

SOLUTION 7-7
Whiteout conditions occur in severe blizzards when the sky and earth seem to blend together, and the horizon vanishes. When these conditions are combined with low visibility so that all familiar landmarks disappear, disorientation can occur. It looks as if you are suspended in a white or gray void. Without the aid of a compass or a precise Global Positioning System (GPS) receiver, you can get lost in your own neighborhood.

Winter into Spring

In the western United States, high pressure often dominates during the winter and spring months, bringing the characteristic clear, cool weather to the southwestern deserts. The northern Rockies and Great Plains get bitterly cold. These conditions result from a strong ridge in the jet stream. The high pressure systems can get intense and become "antistorms." The winds spiral outward from the center, blowing from the west on the polar side and from the east on the equatorial side (Fig. 7-9). Similar situations develop over parts of Europe and Asia.

CHINOOK WINDS

Strong, cold westerly or northwesterly winds, after passing over the mountains at a high altitude, become warmer as they descend into valleys and onto the foothills or prairies. The winds can get gusty and, once in a while, even violent.

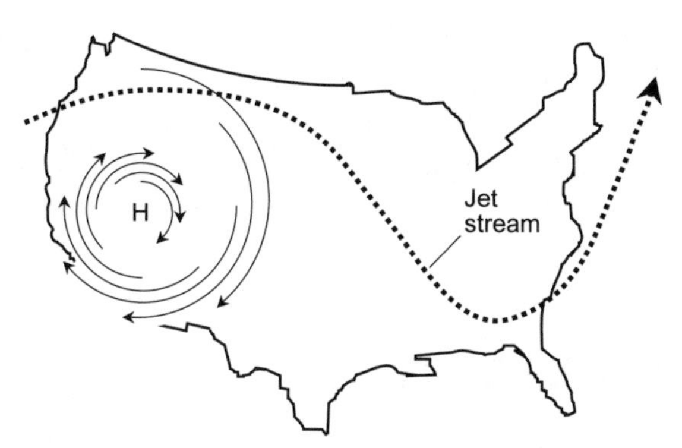

Fig. 7-9. A persistent, strong high-pressure system can cause Chinook and Santa Ana winds in the deserts and mountains of the western United States.

A warm westerly or northwesterly wind, blowing down from the mountains, is called a *Chinook*. Chinooks are most common from January through March. They occur on the north and east sides of high-pressure systems.

Chinook winds routinely cause rapid melting of the snowpack. Temperatures can rise by 20°C (36°F) or more within a few minutes. The warmth and dryness of the Chinook, combined with its high speed, affect the snow like a gigantic electric hair dryer. A Chinook wind provides relief from the bitter cold of winter, but it can also cause damage. Chinook wind gusts in Colorado have been clocked at more than 100 kt (115 mi/h), although the sustained wind speed was considerably less than that. The peculiar gustiness of a Chinook—perhaps 20 kt (23 mi/h) at one moment and 80 kt (92 mi/h) a few seconds later—makes it different from a hurricane, severe thunderstorm, or blizzard. Sometimes Chinooks seem to have a twisting nature, resembling a series of small, invisible tornadoes. Vortex action is occasionally reported by observers as they watch loose dirt or snow blowing around. A Chinook usually occurs without precipitation.

SANTA ANA WINDS

In the south or southwest portion of a wintertime high-pressure system in the western United States, a Chinook-like effect takes place over the mountains in southern California. The primary difference is that the wind is warmer and drier because of the more southerly latitude. As the wind descends into the valleys near the coast, the speed and temperature increase. Gusts of 70 kt (80 mi/h) are not unusual in the Santa Ana area, and for this reason, the wind has been called the Santa Ana wind.

A severe Santa Ana wind event affected large areas of southern California during January 1984, resulting in much damage to trees and buildings. The problem was exacerbated by fires that burned out of control. In desert areas, the Santa Ana wind picked up sand and carried it high into the air. Buildings and motor vehicles got a sand blasting, and people venturing outdoors risked injury.

OTHER MOUNTAIN WINDS

Winds similar to the Chinook and Santa Ana occur frequently in, or near, most mountainous parts of the temperate zone. In central and southern Europe, the Ural Mountains in Russia, and the vast region of western China and Tibet, high-pressure systems sometimes dominate and produce dry, warm, gusty winds. The same thing also happens in the Andes of southern South America.

Mountains produce violent winds in other ways on a local scale. Cool air, descending along a mountain slope, picks up speed like water running downhill. The author experienced this phenomenon in southern California on a night in January. The evening was clear after a magnificent desert day. Then there came a series of howling wind gusts. After a few minutes, it was calm again. Cast-iron patio furniture had been strewn about as if it were made of plastic.

SPRING WINDS

March, like November, is famous for its wind. Winter begins to lose its grip, and people look forward to sunny skies and gentle breezes. The skies are relatively sunny in March in many parts of the world, but the breezes aren't always gentle. The jet stream begins its retreat in earnest by the end of February, and brutal cold waves are not as likely or as frequent as before. It seems that everywhere in the United States, from Maine to California, and from Seattle to Miami, March is a windy month. On a late-March day, the sustained wind can reach gale force despite warm temperatures and sunny skies.

In the American Midwest, Rocky Mountains, and Northeast, March is known for snow as well as wind. Sometimes March brings tremendous snowstorms. Late in the 19th century, a winter storm struck the Northeast with such impact that people tell stories about it to this day. Numerous clubs were organized by storm "veterans" to commemorate the storm of March 12, 1888. Up to 2 m (about 6 ft) of snow fell, and whole houses were buried in drifts.

SPRING FLOODS

Residents of river valleys are acquainted with another feature of late winter and early spring: the floods produced by melting snow. If the winter has been snowy and a few warm days come in March, the result can be a catastrophe. Spring floods carry an extra wallop because the water is dangerously cold, and massive ice floes are carried along by the fast-moving current. These chunks of ice, moving at a few knots, can demolish a house or small building.

FROST HEAVES

Late winter has still another nasty characteristic in places where the ground freezes during the "hard part of the season." When water freezes, it expands. This is not much of a problem in November and December, because in most places the soil is resilient enough to alleviate the pressure as the liquid turns into solid. When the frost melts and the water volume decreases, however, pockets form in the earth, causing damage to paved streets and highways. These are called *frost heaves.*

In March, much of the American Midwest, as well as northern New England, becomes peppered with signs marked "DIP" or "BUMP." Often the signs outlive the danger, but not always! Many accidents occur because people drive too fast over frost heaves. Loose objects can be thrown around inside the vehicle, and if anyone inside the car isn't wearing a seat belt, that person can be hurled against the ceiling. In extreme cases, cars and trucks can go out of control and roll because of road damage resulting from melted ground frost.

MUD SEASON

Severe snows have occurred in April, and even in May, in the northern half of the United States. Snow is a novelty in November. It is pretty at Christmas. It gives us our winter sports, making January and February endurable. In March, most people are tired of it. An April or May snowstorm can be downright depressing!

April can also bring high winds and river flooding, and in some parts of the country, April is the worst month for tornadoes and severe thunderstorms. Although winter can continue well into April, the coldest weather has passed by then, and the precipitation normally falls as rain. If April has a trademark, it is showery weather. After the snow has melted, leaving the ground wet and still frozen in places, the rains provide the final ingredient for a *mud season.*

The upper Midwest, the Great Lakes region, the Ohio Valley, and northern New England have the muddiest Aprils in the United States. The wooded, hilly country of Vermont and New Hampshire is well known (at least by the people who live there) for the early spring mud season. Much of Europe is in the same predicament in April. If the mud season is severe, farmers must delay planting, because tractors get stuck in the mud. In the worst cases, roads get washed out, and in hilly regions, mud slides can become a serious threat.

April does not always bring raw, wet, chilly weather. Sometimes the month is dry and hot. If the preceding winter has been dry, and the rains of April do not come, a severe summer drought may follow. In some parts of the country, this sets the stage for another sort of natural disaster: forest fires.

PROBLEM 7-8

Why are winter temperatures in some high-altitude mountain towns, such as Lead in the Black Hills in South Dakota (alt. 1600 m or 5280 ft), often higher than temperatures in nearby towns such as Rapid City that are hundreds of meters lower in elevation?

SOLUTION 7-8

Events such as this occur in many places with irregular terrain when there is a *temperature inversion.* In regions where the surface is relatively level, inversions also occur, but they are not noticed by most residents because they cannot experience a dramatic change in altitude by driving a few kilometers. Inversions typically occur near frontal boundaries, where cold air pushes under warmer air (cold front) or warm air rises up over the top of a cooler air mass (warm front or stationary front).

PROBLEM 7-9

Once a friend told me she saw it snowing outside when the temperature was 18°C (50°F), far above the freezing point. What can cause this? Was she joking?

SOLUTION 7-9

If there is an "anti-inversion" near the surface in which a cold air mass overlies a warm one, frozen precipitation may not have time to melt before it reaches the ground. The normal tendency of the air temper-

ature to decrease with increasing altitude is exaggerated. This phenomenon, ironically, often occurs in the same sorts of places where inversions occur.

Quiz

This is an "open book" quiz. You may refer to the text in this chapter. A good score is 8 correct. Answers are in the back of the book.

1. The danger of an avalanche on a mountain slope is highest
   (a) when snow on the slope is dry, and has recently fallen in a thin, single layer.
   (b) when snow on the slope is layered, with well-defined boundaries between layers.
   (c) when the jet stream has been further north than normal for an extended period.
   (d) just after an ice storm if there was no snow on the ground to begin with.

2. Refer to Fig. 7-10. This winter weather scenario would be expected to produce
   (a) fair, warm weather in the western United States, and storms in the eastern United States.
   (b) fair, warm weather in the southeastern United States, and cool or cold weather in the western United States.
   (c) fair, warm weather in the central United States, and storms in the eastern and western United States.
   (d) cold, foul weather in the southeastern United States and windy weather in the northwestern United States.

3. In the winter weather situation illustrated by Fig. 7-10, the wind in the Florida panhandle is blowing generally from the
   (a) southwest toward the northeast.
   (b) south toward the north.
   (c) northeast toward the southwest.
   (d) northwest toward the southeast.

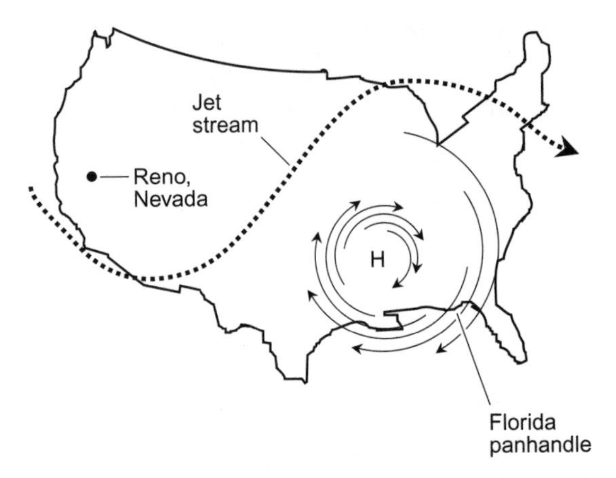

Fig. 7-10. Illustration for Quiz Questions 2 through 4.

4. In the winter weather situation illustrated by Fig. 7-10, we can expect that the weather in Reno, Nevada is
   (a) cool or cold.
   (b) warm and cloudy.
   (c) warm and sunny.
   (d) warm and stormy.

5. Suppose that a winter storm dropped 5 cm (2 in) of precipitation in melted form (that is, if it had all fallen as rain). But the temperature was near freezing, resulting in 25 cm (10 in) of wet snow. If the temperature had been much colder and the snow had thus been much drier and fluffier, which of the following figures represents a good estimate of the snow that would have accumulated from the same storm at the same location?
   (a) 2.5 cm (1 in)
   (b) 10 cm (4 in)
   (c) 25 cm (10 in)
   (d) 50 cm (20 in)

6. High-calorie food can be a good thing in cold weather because
   (a) it keeps you from getting hyperactive.
   (b) it keep you from falling asleep.
   (c) it helps the body to stay warm.
   (d) Forget it! High-calorie food is deadly no matter what.

7. Seattle, Washington is at roughly the same latitude as the extreme northern part of the state of Montana. But Seattle has much warmer winters than any part of Montana. The reason for this is the fact that
   (a) Seattle receives most of its winter storms from the mountains to its east, while Montana receives its winter storms from the cold plains to its east.
   (b) Seattle is dominated by atmospheric high-pressure systems in the winter, while Montana is dominated by low-pressure systems.
   (c) Seattle is dominated by atmospheric low-pressure systems in the winter, while Montana is dominated by high-pressure systems.
   (d) Seattle receives its winter weather systems from the relatively warm Pacific Ocean, while Montana receives its winter weather systems after they have passed over the colder land mass.

8. When the temperature of the surface is below freezing while the temperature of the air is above freezing, liquid precipitation
   (a) often freezes when it lands.
   (b) often falls in the form of hail.
   (c) often falls in the form of snow.
   (d) often causes dangerous downbursts of wind.

9. Severe winter storms do not often make the news in the southern hemisphere because
   (a) severe weather does not often occur in the southern hemisphere, especially in the winter.
   (b) there is not much land mass at the affected latitudes, so not many people experience the effects of winter storms in the southern hemisphere.
   (c) the Antarctic land mass consumes nearly all the available winter storm energy in the southern hemisphere.
   (d) the southern hemisphere has no Hadley cells.

10. Heavy, wet snow can be a cause for concern if it accumulates to great depth because
   (a) it can damage trees and power lines.
   (b) it can block roads and make snow removal difficult.
   (c) it can cause roofs to collapse.
   (d) All of the above

Abnormal Weather

History is replete with examples of short-term and long-term weather abnormalities. The drought period of the 1930s is well remembered even today. Floods have ravaged humankind since the beginning of recorded history. Heat waves and cold snaps can still be dangerous, even in the world's most advanced nations.

Heat Waves

The morning threatens a "scorcher," a copy of the previous several days. By 8:00 A.M. local time, the temperature is uncomfortable. By late morning you sweat, and there is not a cloud in the sky. It's too hot to do anything except swim, but there are no clean lakes nearby. They all contain dangerous chemicals or bacteria. The swimming pools don't open until noon, and they will be overcrowded. No golf, no tennis, no outdoor sports are practical. At 2:00 P.M., the electricity goes off. Too many electric air conditioners are running, and the electric company is blacking out selected areas in a rotating sequence. Perhaps we can fill up the little "kiddie pool" we bought on impulse last fall. We find it, inflate it, and turn on the garden hose. The water pressure is low. Everyone else has the same idea!

LOCATIONS

Heat waves have always been a problem for inland regions throughout the tropics and temperate zones. In the United States, heat waves can develop everywhere. The Pacific northwest, right along the coast, and northern New England are the least vulnerable. The most notorious heat waves occur in the Great Plains.

In the civilized world, forested and mountainous areas are less severely affected than prairies, steppes, and deserts. The far northern and southern regions, such as northern Canada and extreme southern South America, rarely see deadly heat. India and Pakistan are particularly hard hit by heat waves just before the monsoons commence. Certain interior regions of Africa, Asia, South America, and Australia are regularly visited by these events.

CAUSES

The principal cause of a heat wave in the United States is a persistent ridge in the jet stream (Figs. 8-1, 8-2, and 8-3). The worst heat waves occur in summer (obviously), but unusually warm weather can develop at any time of year. A heat wave near the coastline where prevailing winds come off the ocean is normally less intense than the effects of a ridge in or near the middle of a continent, or near the coastline where the prevailing winds are offshore. A pattern such as that shown in Fig. 8-1 causes extreme temperatures in the West. If the jet stream gets oriented as shown in Fig. 8-2, the Midwest has a heat wave. The pattern in Fig. 8-3 results in hot weather along the Eastern Seaboard.

Heat waves nearly always occur in conjunction with high-pressure systems. Part of the reason for this is the fact that high-pressure systems tend to produce

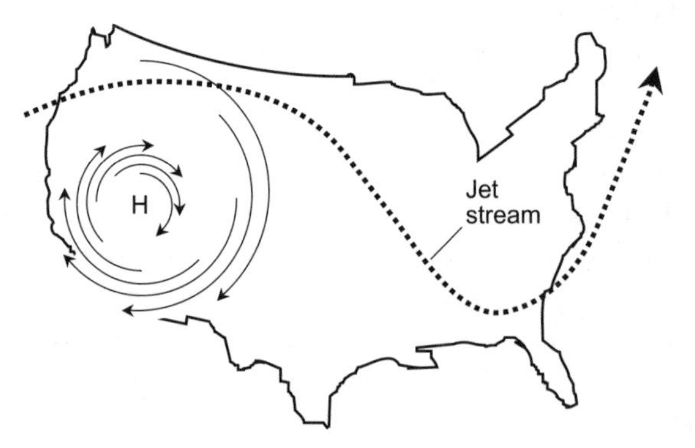

Fig. 8-1. This weather pattern, if it occurs during the summer, will produce a heat wave in the western United States.

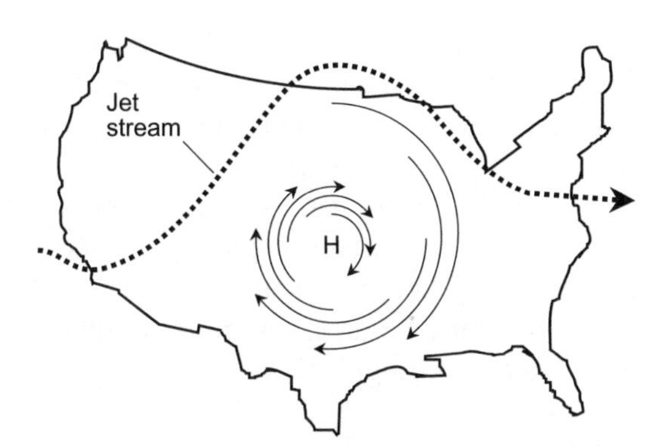

Fig. 8-2. This type of weather pattern is associated with summer heat waves in the central United States.

clear, stable weather. This allows the sun to heat up the surface, which in turn heats the lower atmosphere. In addition, frontal systems are "blocked" by ridges and high-pressure systems. Storm systems tend to pass around the polar edge of the high, following the jet stream, leaving regions on the equatorial side the ridge unaffected by the boundaries between air masses that are normally responsible for precipitation and cooling weather. Without any change in atmospheric circulation, the air gets increasingly hotter and drier until the energy radiated into space is the same as the energy received from the sun. Under persistent clear skies and calm conditions, this occurs at a higher-than-normal temperature.

Heat waves can be rainless, but can nevertheless exhibit high levels of relative humidity, particularly in the southeastern and northeastern parts of the

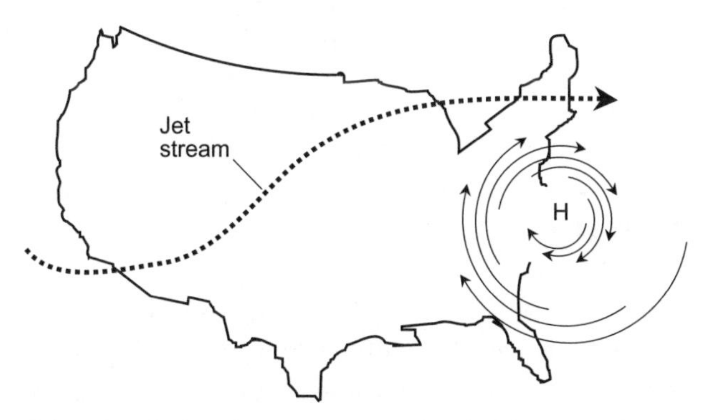

Fig. 8-3. A typical weather pattern associated with summer heat waves in the eastern United States.

United States. A system such as that shown in Fig. 8-3 brings moist air from the Gulf of Mexico over the southeast and east-central United States. The Eastern Seaboard has the muggiest heat waves when the Bermuda high-pressure system fans the continent with moist Atlantic and Gulf air. The circulation on the western side of the Bermuda high is generally from the south, southwest, or southeast. The immediate coast is spared because of the moderating effect of the water, but the air becomes very hot over interior regions. Because of the shape of the east coastline of the United States, the more northerly sections, such as New Jersey, New York, and New England, are often hit the hardest.

THE HUMAN FACTOR

Heat waves can be exacerbated by the actions of governments and corporations, but individual behavior on a large scale also plays a role. In some regions, especially in the tropics in some less developed countries, many of the trees have been cut down. A tree is an effective air conditioner. Bare soil, concrete, and blacktop are excellent heating devices when they are exposed to sunlight. Large cities, therefore, get hotter than rural countryside when cloud cover is reduced.

In major urban areas, hot weather is frequently accompanied by air pollution. The high temperatures and sunshine cause some of the oxygen atoms in the air to group into triplets (O_3) rather than the usual pairs (O_2). The result is *ozone pollution.* This effect is intensified by the components in automobile exhaust. Ozone gives the air a characteristic chlorine-bleach smell. If the concentration is high enough, it irritates the eyes, throat, and lungs. Other pollutants accumulate because of the stability of the high-pressure system, which keeps frontal cyclones from bringing in fresh air.

A severe heat wave is not only uncomfortable, but dangerous. Meteorologists define a heat wave according to the maximum expected temperature for the day. Anything over body temperature (37°C or 98.6°F) makes it official. In some parts of the United States, high temperatures can exceed this level for periods of days or weeks during an especially hot summer.

THE EFFECT OF HUMIDITY

Hot weather is more uncomfortable when the humidity is high, compared to when it is low. This is because the human body is cooled by the evaporation of sweat, and water evaporates less readily as the humidity rises, given constant temperature.

When the humidity is low, the apparent temperature is lower than the actual temperature. When our bodies aren't cooled effectively by sweat evaporation, we really do get hotter. But, although our body cooling systems work best when the humidity is low, it is not necessarily true that we will be comfortable in dry weather. If the air temperature is much higher than body temperature and the humidity is only 10%, you'll still feel hot, as any permanent resident of southern Arizona will tell you. "Dry heat" can, in fact, be more dangerous than "muggy heat," because people are less aware of how high the temperature really is. This can lead to a sense of overconfidence, with potentially serious consequences.

COPING WITH HEAT

Unpleasant effects of hot weather include dehydration, cramps, exhaustion, drowsiness, irritability, and depression. In hot weather, it is important to stay hydrated. This is especially true for children, who often run around with complete disregard for the weather. An easy way to see if you are dehydrated is to weigh yourself at frequent intervals. A liter of water weighs a kilogram; a pint weighs about a pound. Thus, for example, if you suddenly lose 4 pounds, you have lost 2 quarts (about 2 liters) of water. That might not be serious if you are a linebacker for a professional football team, but for a 5-year-old child it represents a significant water loss.

The best way to hydrate is to drink a little water at frequent intervals. Other liquids can be substituted for water, but they might not relieve thirst as effectively as water, and some beverages have side effects. Beverages containing caffeine or theobromine, such as iced coffee or tea, should not be consumed in large quantities because they have a diuretic effect that can worsen dehydration and cause mineral loss, besides producing agitation and heart palpitations in some people. Beer is alright in moderation, but large amounts of alcohol hinder the body's ability to regulate its temperature. "Hard" drinks must be avoided because they can quickly and dramatically worsen dehydration. Pure orange, grapefruit, or tomato juice, as well as other fruit and vegetable juices, are excellent, as long as they aren't loaded with sugar.

Some athletes and coaches believe that salt tablets can help prevent exhaustion and cramps in hot weather. Others advise against taking them, and insist that it is potassium loss, rather than sodium loss, that causes the most trouble, and that you're better off eating a banana than taking a salt tablet. Consult your doctor concerning the advisability of taking any type of nutritional or chemical supplement to help you deal with the heat.

Foods loaded with sugar or fat, or large quantities of food, should be avoided in extremely hot weather. Such foods are useful in cold weather because they can help warm you by fueling your metabolism, but in hot weather this is an undesirable effect. A large meal can put extra stress on an already overtaxed body. Moderate meals at frequent intervals are better than large meals spaced far apart in time. Light foods such as salads and fruits are better than heavier fare such as gigantic steaks. During a heat wave, some people find that their food preferences naturally change in these ways.

The problems of drowsiness, irritability, and depression are difficult to cope with directly. A little (not a lot of) iced coffee or tea can help alleviate these symptoms, as long as your doctor doesn't forbid you to have caffeine. The best thing to do is to try to keep cool, literally and figuratively.

Heatstroke represents the most dangerous health threat during a heat wave. In heatstroke, the body loses the ability to control its core temperature. The result is a high fever of 40.5°C (105°F) or more. Sweating stops, and the victim appears pale or gray. Unconsciousness usually occurs. Unless a heatstroke victim receives first aid and medical attention immediately, he or she may die. For more information, consult up-to-date American Red Cross first-aid publications.

COOLING DEGREE DAYS

People who own air conditioners know that electric bills can escalate during a heat wave. Of course, it is better to run the air conditioner than to die of heatstroke, but the thermostat should be adjusted for the highest tolerable temperature. This conserves energy, and it narrows the margin between indoor and outdoor temperatures, minimizing the "thermal shock" for people going in or out of a building.

The severity of a particular summer, in terms of heat, is measured according to an average figure called *cooling degree days.* In the United States, the Fahrenheit scale is used to calculate cooling degree days. The figure is based on the average temperature relative to 65°F. A single cooling degree day represents one day for which the mean temperature is 66°F. If the average temperature on a given day is 75°F, it represents 10 cooling degree days. The figure is added cumulatively over a period of time. Figure 8-4A is an example of a cooling-degree-day graph for a hypothetical week. Figure 8-4B shows the graph for a whole season (May through September). The slope of the curve indicates the relative energy consumption that is required to keep the indoor environment comfortable.

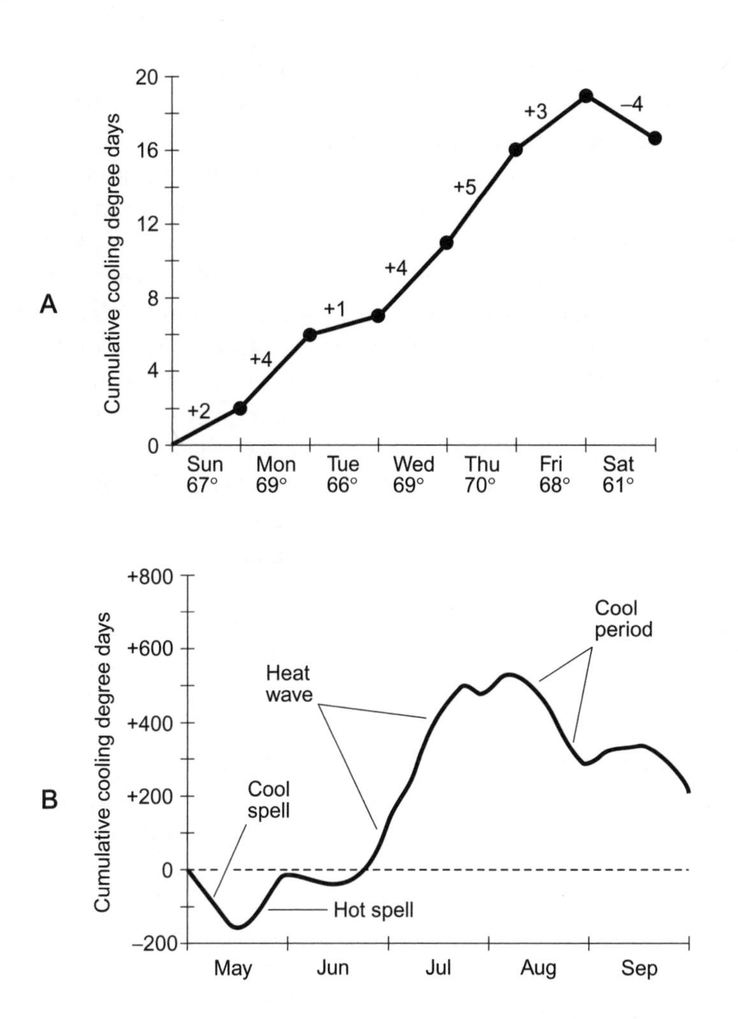

Fig. 8-4. Accumulation of cooling degree days (Fahrenheit) for a hypothetical week (A) and a hypothetical season (B).

HEAT ISLANDS

The temperature at a particular surface location is affected not only by natural weather factors, but by the nature of the terrain, geography, and the type of material that makes up the surface. High in the mountains, severe heat waves never occur. Deadly hot weather is rare near the seashore, on the leeward shore of a very large lake such as Superior, or in areas with vast, dense forests. Severe heat is worsened by open plains, prairies, or steppes, especially if there is not much vegetation. Large cities are the worst of all.

Most of the time, the interior of a city is somewhat warmer than the surrounding countryside. This happens for two reasons. First, concrete, asphalt, and other human-made materials do not cool off by evaporation as effectively as vegetation or soil, because they do not absorb much water to begin with. Second, there are sources of heat within a city, such as automobiles, electric lights, and power plants, that are much less concentrated in rural places. In the downtown area of a large city, the temperature can be several degrees Celsius higher than in the countryside a few miles away. The highest temperatures are observed near the center of a city. The heat zone in a city is known as a *heat island.*

Heat islands are of obvious benefit in extremely cold weather. The expense of heating during the winter is generally lower in a metropolitan area than in rural areas, but in the summer during a heat wave, the heat island makes the situation worse. Because of this, urban people flock to rural areas during heat waves.

EL NIÑO

Global weather patterns are influenced by factors we do not completely understand. Hot and cold spells, fluctuations in precipitation, and frequency of storms are affected by ocean temperatures. One of the most interesting, noticed for hundreds of years by natives of the eastern equatorial Pacific, is a strange rise in the water temperature. It occurs at unpredictable intervals, lasting from 1 to 3 years. This oceanic heat wave, given the Spanish name *El Niño* (The Child), is associated with climatic variability around the world.

An El Niño event is actually a combination of atmospheric and oceanic phenomena. Semipermanent areas of atmospheric high and low pressure over the Pacific reverse their positions, and the ocean temperature in some locations, especially off the coast of Peru, rises by several degrees Celsius. Atmospheric winds and ocean currents change course, bringing wet weather in some areas and drought to others. Floods, triggered by tropical storms and excessive rainfall, batter Peru and Ecuador in South America, parts of the west coast of North America, and the southeastern United States. Dry weather causes heat waves, dust storms, and crop losses in Mexico, southern Africa, Australia, and Indonesia. In southern Asia, the seasonal monsoon rains may fail.

Normally, high atmospheric pressure prevails in the eastern Pacific, and low pressure dominates the western Pacific. This is why the western mountain slopes of South America are chronically dry, while Indonesia is covered by rain forests.

Pacific Ocean water temperatures are normally warm in the west and cool in the east. In an El Niño event, this pattern reverses. A massive low-pressure atmospheric system develops over the eastern Pacific, and a high-pressure area forms over the western Pacific. The easterly winds, normally prevalent in the tropics, slacken or reverse. The cold water that usually rises up from the deep ocean off the coast of South America no longer rises, and the surface water temperature therefore increases.

El Niño events can have beneficial effects in some regions. Although the frequency of tropical cyclones may increase in the western Pacific because of the low pressure off the coast of South America, there are fewer hurricanes than usual in the Atlantic Ocean, the Caribbean Sea, and the Gulf of Mexico. High-level air currents, generated by the effects of El Niño, move across Central America and over the storm spawning grounds in the Atlantic, Caribbean, and Gulf. These upper-atmospheric winds shear off the tops of some tropical storms before they can develop into major hurricanes. (This does not, however, mean that a severe Atlantic, Caribbean, or Gulf hurricane cannot develop in an El Niño year.)

El Niño events are correlated with dramatic climate fluctuations all around the world. Similarly, the opposite of an El Niño event (called *La Niña*) can have worldwide effects. La Niña occurs when the high-pressure atmospheric system over the eastern equatorial Pacific strengthens, the easterly tropical winds blow harder, and more cold water than usual wells up from the deep ocean off the coast of South America. It has been suggested that the unusually severe winter of 1983–1984 on the North American continent was triggered, in part, by a La Niña event. Active Atlantic hurricane seasons can also occur in conjunction with La Niña.

PROBLEM 8-1

Ozone is supposed to protect humans and other animals from the damaging ultraviolet (UV) rays from the sun. Doesn't this suggest that ozone pollution over large cities might actually be a good thing?

SOLUTION 8-1

The protection from excessive UV that has been attributed to ozone is the result of the natural presence of these molecules in the upper atmosphere. This is a large-scale phenomenon, affecting the entire surface of the earth. Ozone pollution in cities occurs only near the surface, and is a localized effect. In theory, this ozone might reduce the amount of UV to which city people are exposed, but this potential benefit is more than offset by harmful effects, especially for people with respiratory problems.

Cold Spells

At the opposite end of the temperature spectrum, prolonged cold spells cause as much trouble as heat waves. During some winter seasons, the polar air mass advances far toward the equator over the North American continent and stays there for weeks, producing subnormal temperatures, and often above-normal precipitation. Whenever and wherever a wintertime jet-stream trough becomes fixed in place, a severe winter is the result. Weather events of this type are common in Europe and Russia as well.

THE NORTH PACIFIC HIGH

A deep trough in the jet stream over North America often occurs when the North Pacific high-pressure system intensifies. The clockwise circulation around the northeastern part of the offshore high pumps arctic air into the continent. Depending on exactly where the high-pressure region is centered, how large it is, and how strong it is, the cold wave may affect primarily the midwestern United States, as shown in Fig. 8-5A, or the Ohio Valley and East Coast, as shown in Fig. 8-5B. A situation such as that shown at B can sometimes cause freezing overnight temperatures as far south as Miami, Florida.

Cold waves in some parts of the country are often accompanied by warm weather farther west, because the Pacific high represents a ridge in the jet stream. For example, a brutal cold wave in the Dakotas might occur along with warm, sunny weather in California and Arizona. For those residents of the desert Southwest whose livelihood depends on tourism, such events are welcomed. Skiers and snowmobilers in the Black Hills, ironically, may welcome it too, because the trough is likely to be preceded or accompanied by heavy snow.

A similar jet stream trough can develop over eastern Europe in conjunction with an intensification of the Bermuda high. Spain and Portugal then have balmy, fair weather, while rains drench Norway, Sweden, Holland, and Poland, and bitter-cold storms sweep across the Russian plains. In the southern hemisphere, the effects of jet stream troughs are less brutal in terms of cold temperatures, but the tip of South America is frequently blasted by raw, wet gales as storms sweep around troughs in that region.

MODERATING FACTORS

Cold waves, like heat waves, are moderated by the presence of the ocean. In England, cold waves such as those in the midwestern and north central United

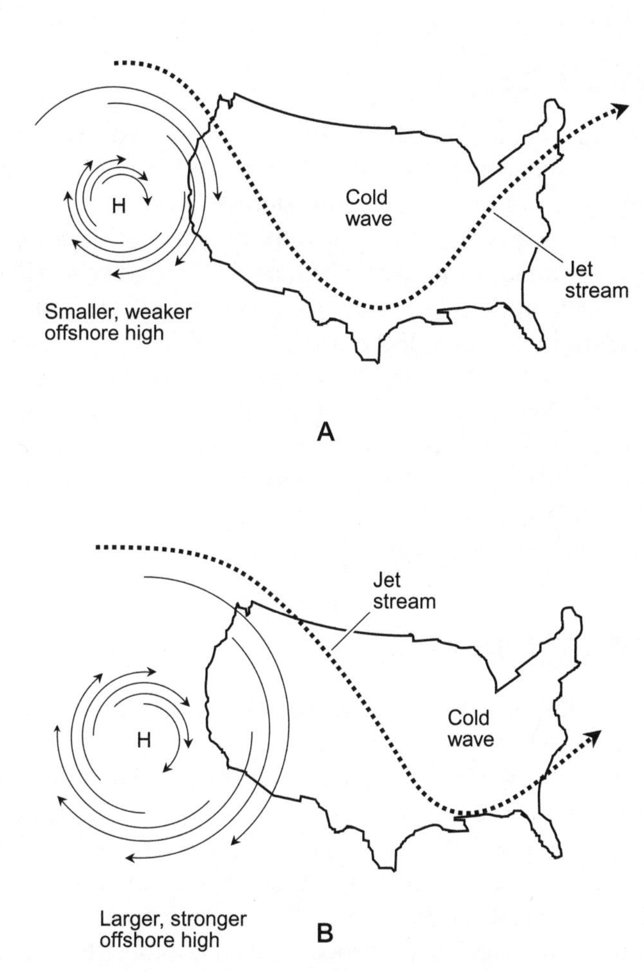

Fig. 8-5. A Pacific high can cause a cold wave in the central (A) or eastern (B) region of the United States, depending on the size, location, and intensity of the high-pressure system.

States are unknown, despite the fact that England is further north in latitude. This is because of the moderating effect of the warm Atlantic current that originates in the Gulf of Mexico.

In Anchorage, Alaska, winter weather is usually warmer than that at comparable latitudes inland, because of the proximity of the warm North Pacific current. On some winter night, get on the Internet and go to the Weather Channel (www.weather.com). Compare the wintertime temperatures at Anchorage, Alaska, and an inland place at approximately the same latitude, such as Yellowknife, in the Northwest Territories of Canada.

COLD WAVE ZONES

Because jet stream troughs frequently develop and persist over inland parts of large continents in the winter, conditions often become favorable for severe, prolonged cold waves in the upper-midwestern United States, central Canada, and eastern Europe. A less populated, but still famous, haven for cold waves is Siberia, in the eastern part of Russia. Much of Siberia, however, lies within the arctic circle, and is therefore not a true part of the temperate zone. Cold winter weather is the norm in that part of the world.

Cold waves can make human affairs difficult in many ways. All but the most hardy outdoor sports enthusiasts must suspend their activities when the temperature plummets to frigid levels. A greater concern is the high cost of keeping warm. Energy bills rise to astronomical levels during severe cold snaps. This is in part because some heating systems can't keep up with unnaturally cold weather, and in part because the price per energy unit rises as a result of increased demand.

HEATING DEGREE DAYS

The severity of a cold season is determined in much the same way as the severity of a hot season. Relative demand for energy during a winter is calculated according to the accumulation of *heating degree days.* The basis for determining heating degree days is a mean temperature of 65°F. A single heating degree day represents a day for which the average temperature is 64°F. If the mean temperature is 55°F for 1 day, we have 10 heating degree days. Fig. 8-6A shows an example of the accumulation of heating degree days for a hypothetical week. Fig. 8-6B illustrates an entire hypothetical season (October through April). The most severe cold waves correspond to regions in the graph at B having a steep downward slope. The slope of the curve is an indication of the amount of energy required to keep the indoor environment comfortable.

In most places at temperate latitudes, the number of heating degree days is greater each year than the number of cooling degree days. In New England, for example, the number of cooling degree days is normally a few hundred for each warm season, but the number of heating degree days is several thousand. The same holds for most of the United States, with the exception of Hawaii, Florida, the desert Southwest, and the extreme West Coast.

ALTERNATIVE ENERGY FOR HEATING

In recent decades, the cost of energy has been increasing rapidly, mostly because of the dramatic rise in the price of petroleum. Alternative energy sources are

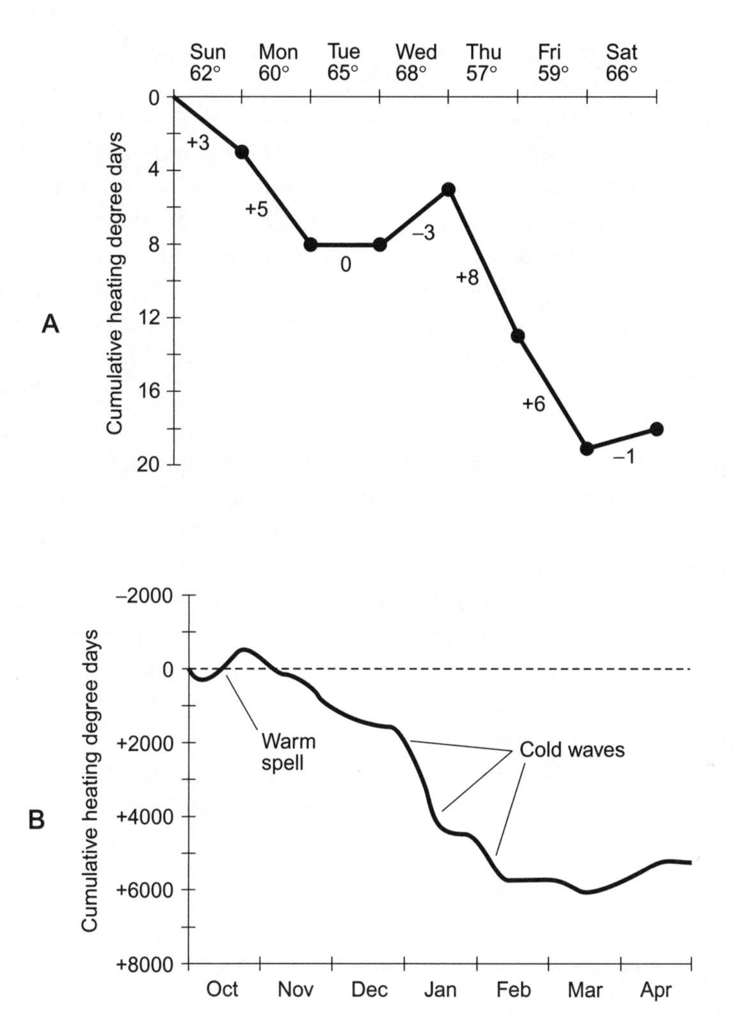

Fig. 8-6. Accumulation of heating degree days (Fahrenheit) for a hypothetical week (A) and a hypothetical season (B).

becoming more attractive, especially for heating purposes. People have to stay reasonably warm to survive. This problem will get more attention in future years, as some forms of energy become unavailable or unaffordable.

The sun is one of the most promising alternative energy sources. In some places, solar energy is a practical and effective source of heat for small buildings. The best locations are, not surprisingly, those at which the sun shines much of the time. In general, the weather is sunnier in the western United States than in the east, and the sunlight is more intense in the southern part of the country than in the northern part. Solar energy, however, can be used to a limited extent even in regions where cloudy weather is common.

Solar energy can be harnessed directly by placing large windows on the south side of a building, in conjunction with heat-absorbing and heat-retaining floors, walls, and furniture. When the sun shines through the windows, visible light and shortwave infrared (IR) penetrate the glass. Objects in the room get heated and radiate longwave IR, to which the glass is opaque (Fig. 8-7). The longwave IR continues to be radiated from objects in the room after the sun has gone down or the weather has turned overcast. Direct solar heating of this sort, also called *passive solar heating,* can be effective anywhere the sun shines on most days, even if the outside temperature is frigid during part of the year. The Bighorn Basin of Wyoming is a good example of a region in which passive solar heating is practical, even though wintertime temperatures can remain below freezing for days or weeks at a time. Passive solar heating would be less practical in a region such as the coast of Oregon, where, although temperatures are not terribly cold in the winter, the sky is overcast almost every day, all day.

In less sunny locations, indirect solar heating can be used. An indirect system uses solar panels to generate electricity for heating, and also for general power needs. Such a system can be independent of the utility company (a *stand-alone solar-electric system*), or it can be interconnected with the commercial power line (an *interactive solar-electric system*).

The wind can be harnessed to provide heat by generating electricity. Farmers have been using the wind as a source of energy for hundreds of years. A single

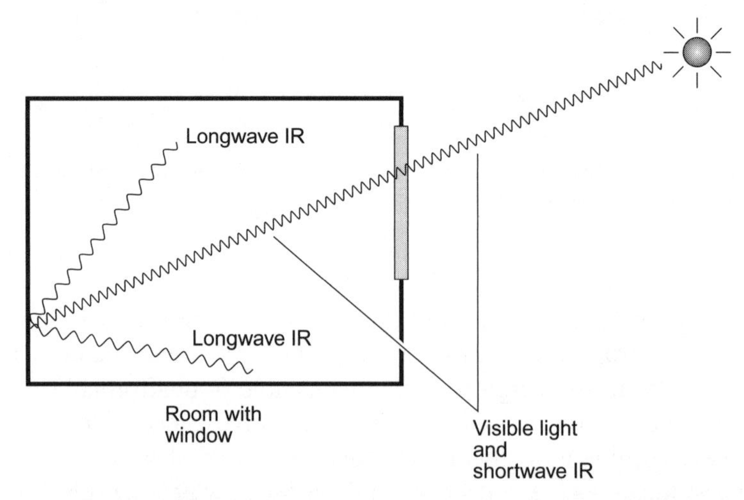

Fig. 8-7. When sunlight shines through a window into an enclosed room, shortwave IR and visible light penetrate the window glass, but the longwave IR does not. Thus, heat energy accumulates in the room.

wind-driven generator can produce several kilowatts of power. Such generators are available from commercial sources.

Other means of keeping warm, all of which will be considered in the coming decades, include wood stoves, pellet stoves, fireplaces (if they are efficiently designed), geothermal heat, and nuclear power. Natural gas, although a petroleum product, may someday be replaced by hydrogen derived from electrolysis or from chemical reactions. Some scientists have even suggested that massive reflectors could be put in space to direct solar energy to earth-based power plants.

Energy conservation is important if a house or building is to be kept warm at an affordable price. Heat loss can occur because of *radiation, conduction,* or *convection.* Radiation loss can be detected by photographing a structure at IR wavelengths. Conduction loss can, in some cases, also be detected in this way. Convection contributes to heat loss when a building is not airtight enough. If you place your hand at the base of your front door on a blustery day and feel cool air currents, you have significant convection heat loss. However, a home that is too airtight can be problematic, because the ventilation may not be sufficient for good health. A professional should be consulted regarding the insulation of a particular house or building.

PROBLEM 8-2

A long time ago, I was reading an issue of the *Farmer's Almanac,* in which the year 1816 was described as "the year without a summer." Was this story made up? If it really happened, what could have caused it?

SOLUTION 8-2

The "cold summer of 1816" is well documented. Snow and frost occurred in June, July, and August throughout much of Europe and the United States. Meteorologists and climatologists believe that the cause was the explosive April 1815 eruption of Mount Tambora in Indonesia. The volcano threw huge quantities of dust into the atmosphere up to altitudes of 40 km (25 mi). It took more than a year for this dust to settle out. In the meantime, it reduced the amount of energy the earth's surface received from the sun. The most noticeable effect was a cooler-than-normal summer throughout the northern hemisphere in 1816. Similar events can be expected from time to time in the future.

PROBLEM 8-3

Why does installing foam insulation in the walls of a house reduce the heat loss? Isn't "dead air" space between the inner and outer walls just as good as foam insulation?

SOLUTION 8-3

The answer to this lies in the fact that the "dead air" between the inner and outer walls in an uninsulated building is not really "dead"! It carries heat away by means of convection. The difference in temperature between the inner and outer walls sets up air currents that act to transfer heat from the warmer surface (the inner wall) to the cooler one (the outer wall). The same is true in an attic that lacks insulation. Foam insulation chokes off these convective currents. Fiberglass works well also, although some precautions must be taken when working with it.

Too Much Water

Floods are responsible for millions of dollars of damage and dozens of deaths every year in the United States. In some countries, the human toll can be much higher. Most people who live on river flood plains know what can happen, even if they haven't been there long enough to see a catastrophe. Floods can occur over a prolonged period of heavier-than-normal precipitation. They can also occur suddenly as a result of a single storm.

FLASH FLOODS

In July 1978, following a wet spring in the north-central United States, a line of heavy thundershowers developed in southern Minnesota along a stationary front oriented in an east–west direction. Weather reports warned of imminent heavy rain. Cumulonimbus clouds moved toward the north and bore down on the city of Rochester and the surrounding area. Rain began, driven by a southeasterly gale. Most heavy rain showers in this region last for 20 to 40 minutes, but the July 1978 storm continued unabated for hours.

At about 11:00 P.M. Central Daylight Time (CDT), the Zumbro River had grown from a quiet little stream to a waterway out of control. Within its newly claimed banks lay about 30% of the city of Rochester. Residents were awakened by evacuation orders, and some were surprised to find knee-deep water in their basements and garages. By midnight, about 18 cm (7 in) of rain had fallen. It was the worst flood in the history of the city up to that time. Route 63, which runs north and south and forms the main avenue (Broadway), looked like a boat launching ramp in the vicinity of 16th Street N.E. This was a classic example of a *flash flood.* The term derives from the fact that such floods occur quickly, and sometimes almost instantly ("in a flash").

Flash floods are especially dangerous because of the rapidity with which they strike. In the Rochester flood, the water level rose so fast that its vertical progress was plainly visible. When water rises that fast, strong horizontal currents are produced. Even water that looks calm may be flowing at several knots. Floating debris becomes a deadly hazard.

Flash floods can occur in desert areas. Las Vegas provides an excellent example. If there is a rain storm that produces only 5 cm (2 in) of rain in a few hours, a flood is likely because the water runs off rather than soaking in. Even the Sahara gets an occasional deluge, and small lakes can form if the rain accumulation is significant. Such floods can cause problems for people living in or near the area, but, after a short time, an ephemeral desert flood lake evaporates, leaving things as they were.

TRIBUTARIES AND FLOOD PLAINS

Certain types of rivers and streams are more susceptible to flooding than others. Floods are most common in regions where the rainfall varies significantly. The most flood-prone rivers and streams are those with many tributaries or which run through a flat plain.

In India, the Ganges River regularly floods during the monsoon season because a large part of the country gets its drainage by means of tributaries. A heavy rain anywhere upstream, in any of the tributaries, is followed by a rise in the water level downstream. The people who dwell in the narrow flood plain of the Ganges are used to the annual inundation, and treat it as an ordinary event. In recent years, some measures have been taken to reduce the damaging effects of flooding along this river.

The Yangtze in China is another river that is susceptible to regular and major flooding. Because the flood plain is so low, the water level does not have to rise very much in order to put a large land area under water. But a big storm can bring huge amounts of extra water into the Yangtze. As a result, crests can sometimes reach 30 m (about 100 ft) above the normal water level!

Several of the world's rivers are famous for the beneficial effects of their flooding; an example is the Nile in Egypt. For those who dwell next to the Nile today, the flooding is less dramatic than it was in earlier times. In a country largely covered by deserts, the annual Nile flood was, for thousands of years, a vital link in the food chain. Now a dam controls the waters of the Nile, taming the destructive effects of its irregular floods, but also upsetting the natural fertilizing process that used to occur in the river valley.

DAMS

Devastating floods can occur downstream from large dams. Ironically, one of the primary purposes for building a dam is to control flooding. The Aswan High Dam in Egypt was built to regulate the level of the Nile, reducing flood damage in the valleys and providing water for irrigation. Behind the dam, a massive reservoir has formed. The Aswan High Dam, completed in 1970, is one of the largest in the world. If the dam were to break, the colossal reservoir known as Lake Nasser would spill down the Nile to the Mediterranean Sea. It would be like a tsunami on the Nile! Nevertheless, the benefits of this dam are considered to far exceed the risks.

A grisly example of what can happen to a dammed river took place in the Johnstown, Pennsylvania, flood of May, 1889. A large earthen structure called the South Fork Dam held back a reservoir about 3.2 km (2 mi) long, on the shores of which a resort complex was built. The dam was inadequately maintained, and a rainstorm caused it to disintegrate. A wall of water cascaded down the river, carrying away everything in its path. Johnstown, a few kilometers downstream from the dam, was obliterated, and several other communities suffered the same fate.

Earthen dams are the most common type, primarily because they are less expensive than concrete dams. They are also more easily destroyed by a phenomenon known as *overtopping,* in which a swollen reservoir overflows the dam, eroding it away almost instantly. Proper dam construction prevents overtopping and minimizes the danger of a catastrophe such as the Johnstown flood—but no dam is indestructible. A big enough storm can destroy any dam.

FLOOD-PRODUCING WEATHER

Two types of weather phenomena can produce river flooding: excessive rainfall and rapid snowmelt. In both cases, the situation can build up gradually over several weeks or months, exploding into a flood because of a single rainstorm or warm spell. Sometimes both of these weather factors act together to produce a "maxi flood."

Heavy rains are associated with intense atmospheric low-pressure systems in spring and summer. As a warm or cold front progresses around the center of a temperate-latitude cyclone, the front slows down and may stall. Fronts tend to slow down or stall as they become aligned in the direction of movement of the cyclone. Thus, instead of sweeping past points in its path, the front hovers in about the same position for days. Figure 8-8 shows a situation of this kind that

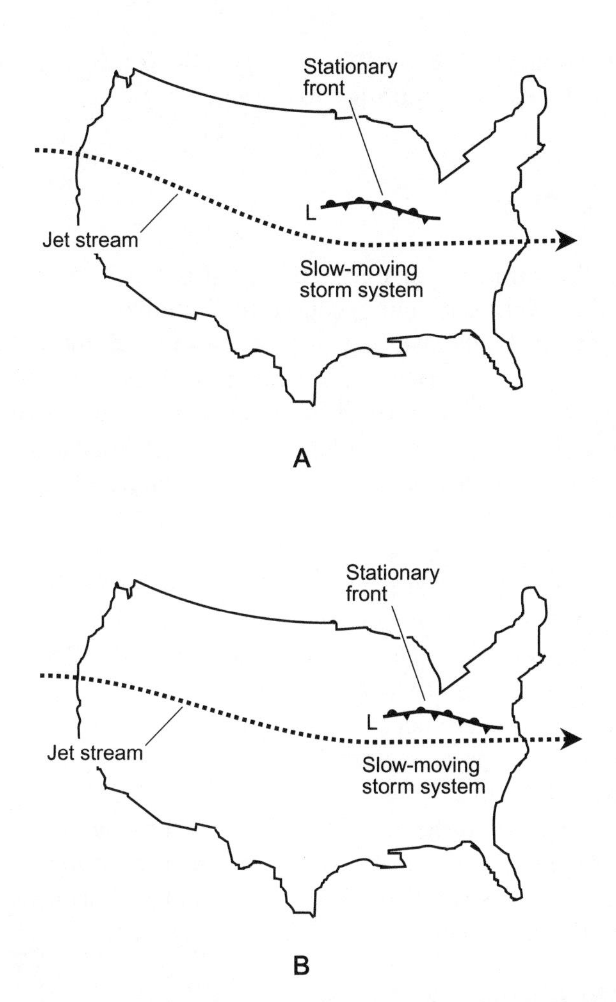

Fig. 8-8. A typical flood-producing weather scenario. The low-pressure system, and its stationary front, move slowly. The maps at A and B represent conditions 24 hours apart.

would generate a series of rainy days in the Ohio Valley. In some low-pressure systems, heavy rains occur mainly on the polar side of the center. This situation is particularly true of extratropical hurricanes moving generally from southwest to northeast after they have passed inland from the Gulf of Mexico.

Whenever there is a winter with unusually heavy snow, residents of flood plains fear the arrival of spring. If a warm spell occurs, streams and rivers begin to swell. If the onset of the warm weather is rapid, the ground cannot absorb the runoff because there is too much water and the ground frost has not had time to thaw. Warm spells are frequently followed by cold fronts or stationary fronts, and the associated weather can bring heavy rain.

SAFETY IN A FLOOD

Some floods strike with little or no warning, but others can be predicted hours or days in advance. Spring flooding, caused by the melting of snow, is the easiest to forecast. Flash floods resulting from rainstorms are more difficult to predict, but meteorologists can usually tell when a flash flood is likely. Dam-burst floods are the hardest of all to predict in terms of the time of strike. Because of this, and because of their extreme violence, dam-burst floods, while rarities, are the most dangerous. Here are some common-sense safety rules to follow before and during a flood event:

- If a flood warning is issued, evacuate if advised to do so
- Move valuables to an upper floor if time permits
- Turn off the main electrical switch or breaker if the flood might put any part of your house underwater
- Do not wade anywhere near downed utility lines, because water conducts electricity well enough to cause electrocution even if you do not come into contact with the wires
- Do not try to swim in flood waters, except when unavoidable
- Beware of using a boat in a flood, especially if the water is moving
- After the water has receded, check with local authorities concerning the safety of drinking water

PROBLEM 8-4

Why might a river rise to flood stage even when the local weather has been dry, or the winter has not been snowy?

SOLUTION 8-4

If a storm occurs anywhere over a river or its tributaries, floods can occur for some distance downstream. The drainage system carries the excess water away from the storm-affected region. For example, suppose a stationary front over central Minnesota and Wisconsin produces severe flooding of the tributaries of the Mississippi River in that region (the St. Croix River, the Minnesota River, and the streams that feed them). A couple of days later, a flood may occur in St. Louis, Missouri, even if the weather there has been fine.

Not Enough Water

Throughout history, the prospect of a prolonged *drought* has been feared more than the danger of occasional flooding. It is almost always possible to flee to high ground when it rains too much, but during a drought there is no escape. The entire human community is placed at the mercy of the weather. If crops won't grow and imports are not available or are unaffordable, people can't eat.

In some parts of the world today, drought (also called *drouth*) is slowly starving millions of people. Most of us tend to think of Africa especially, and to some extent Asia, when we imagine severe, prolonged drought, but it could—and someday it might—happen in the industrialized world.

THE DROUGHT BELTS

Some parts of the world have a much drier climate than others. In some cases this is a product of the geography. For example, deserts commonly exist on the leeward sides of mountain ranges. Persistent high pressure centered near 30°N and 30°S latitude has created a state of unending drought in certain places. You can look at a map of the world and see that major deserts are concentrated near these latitudes, where the temperate and torrid zones converge. The Sahara in northern Africa is the most well-known example. Deserts also exist in southern Africa. Progressing eastward around the north 30th parallel, we find the deserts of Saudi Arabia, Iran, Pakistan, Tibet, the southwestern United States, and northern Mexico. In the southern hemisphere, Australia is largely covered by deserts, and the climate of west-central South America is also arid.

Not all regions near 30°N and 30°S latitude have desert climates. Florida and most of the United States Gulf Coast, eastern China, and eastern Australia get a fair amount of rainfall. These regions receive adequate rainfall because moist tropical air blows inland regularly from warm ocean waters.

Places that do not get much precipitation show more variability of precipitation from month to month and year to year than places that get a lot of precipitation. Some deserts get an entire year's rainfall in a single storm. In other places, there is a *wet season* and a *dry season.* Most of the precipitation in southern California comes during the winter months. In India and Pakistan, summer monsoons bring most of the rain for the year.

In recent decades, the world's deserts, even those long considered uninhabitable, have been visited and populated by humans. This is especially true of the southwestern United States. The desert ecosystem is fragile, and human settlement may harm the environment. There is evidence that the world's deserts are expanding, but there is disagreement as to whether that is because of human activity, or because of natural changes in climate. The environment on our planet has changed, is changing, and will continue to change. As recently as 5000 years ago, some scientists believe, much of the Middle East was green and fertile.

Semiarid regions occasionally experience droughts that render them practically deserts. These droughts are disruptive to humankind, because people rely on semiarid places such as the Great Plains for much of the food supply. During the 1930s, a prolonged drought occurred in the central United States, resulting in near famine in some regions. Other droughts, less severe, have taken place since then.

ARE DROUGHTS CYCLIC?

Temporary droughts, such as the one that took place during the 1930s, occur when the high-pressure belt is closer to the pole than normal. The conditions that produce a temporary drought are the same as those that cause a heat wave: a persistent ridge in the jet stream over a continent. Then, in the northern hemisphere, the desert belts shift or expand northward. Summer rainfall decreases while the temperature increases.

There were other droughts during the 20th century besides the "dust bowl" of the 1930s. The periods of 1912–1914, the middle 1950s, the middle and late 1970s, and the late 1990s were characterized by above-normal temperatures and below-normal rainfall over much of the United States. These drought periods were spaced at intervals of a little more than 20 years. Is this a coincidence, or does it indicate a cycle? If droughts are periodic in nature, what factors are responsible for the cycle?

Sunspots have been suggested as a cause of climate cycles (and plenty of other phenomena, too!). The number of sunspots varies from year to year, reaching a maximum approximately once every 11 years. Each succeeding sunspot maximum takes place during a period of opposite solar magnetic polarity. Thus, the full sunspot cycle has a period of about 22 years. Some climatologists think sunspot numbers influence the climate, and recent studies have shown that the radiation output from the sun actually fluctuates along with the sunspot cycle. It would not take much of a change in solar radiation to cause a significant climate change on the earth.

Some scientists are skeptical of the cyclic theory of drought, and especially the sunspot theory, pointing out that humans have a tendency (almost an obsession) to seek cycles in nature. They suggest that people may think a cycle exists for a certain phenomenon when in fact there is none. Even in a chaotic system, occasional repetitive events, similar to cycles, take place. People also sometimes read their own conclusions into observed data. It has been said, only half in jest, that no one should let reality disprove a good theory.

It takes more than a few consecutive events to make a solid case for the sunspot-cycle theory of drought. Unfortunately, the only way to be sure about this is to continue observations for another couple of centuries, at least. Even then, the data will have to be evaluated objectively.

THE EFFECTS OF HUMAN ACTIVITY

During the 20th century, the 1930s drought was much worse than any of the others. The reason is not certain, but perhaps improper farming techniques had something to do with it. People found out that New England farming methods did not work in the Great Plains. When the soil "out west" was plowed, it dried out during the summer months, and the almost constant winds blew it away. After the 1930s, trees were planted as windbreaks, and much of the land was irrigated.

Droughts are correlated with above-normal temperatures. Are there causative factors at work? Do higher temperatures cause droughts, or do droughts cause higher temperatures? Or are both the heat and the drought caused by some other factor, such as above-normal atmospheric pressure over a region, or increased or decreased solar radiation? The exact cause-and-effect relationship is not known, although theories abound.

In recent years, *global warming* (a general upward trend in the average temperature of the earth's atmosphere) has become an issue. If drought is caused by temporary global warming, such as would be produced by an upward spike in solar radiation, there is little we humans can do about it. But evidence has been mounting that human activities are making the earth warmer. Pollutants, especially *carbon dioxide* (CO_2), are believed to be the main culprits.

When visible light and shortwave IR strike the surface, especially a land mass not covered by snow or ice, the earth is warmed and emits longwave IR. Some of the longwave IR is radiated back into space, but some is also trapped by the atmosphere, in a manner similar to that shown in Fig. 8-7, in which solar energy warms up a room with a glass window. This is the same effect that keeps a greenhouse warm in winter. For this reason, the phenomenon is called the *greenhouse effect.* The CO_2 in our atmosphere contributes to this effect. This colorless, odorless gas is present in trace amounts (less than 1% of the air at the surface, by weight). If there were less CO_2, the earth would be cooler, and if there were more, the earth would be warmer, all other factors remaining constant.

The burning of conventional fuels such as wood, coal, and methane (natural gas) produces CO_2 as a byproduct. As more and more people have begun to use these fuels, particularly in developing countries and emerging economies such as China, the amount of CO_2 in the earth's atmosphere has been increasing. The average temperature in the atmosphere has, over the same period, grown warmer. Most scientists today believe that these phenomena are correlated, and that there is a causative effect. The consensus is that the use of conventional fuels is making the earth warmer than it would be if these fuels were not used. Whether the consequences of human-induced global warming will ultimately prove to be good or bad can be debated, but few scientists continue to call it fiction.

As things work out, a small increase in atmospheric CO_2 can make a big difference in the extent to which the earth radiates longwave IR energy back into space. The burning of *fossil fuels* in particular is raising the concentration of CO_2 in our air. Some climatologists have warned that, if fossil fuels continue to be consumed in increasing amounts, the average temperature of the earth 100 years from now might be several degrees Celsius warmer than it would be if alternative fuels were used.

An increase of 5°C (9°F) in the average global temperature would have a profound impact on our climate. A global warming trend could cause the annual precipitation to change in many regions. The zones of the trade winds, prevailing westerlies, and polar easterlies might be redistributed. Suppose a new Hadley cell developed near the equator, producing unbelievably hot deserts? Then previously fertile farmland would turn into desert. Some desert areas, in contrast, would become arable. The level of the oceans would rise because some of the ice now locked up in glaciers and polar ice caps would melt. Many of the world's major coastal cities would be submerged. Drinking water in these places would be contaminated with sea salts.

VERDICT: UNCERTAIN!

Certain purveyors of gloom paint a bleak picture of global warming, condemning all human activities no matter what their purpose or intent. Other people are more pragmatic. Some regions would actually benefit from a rise in the average temperature of the earth. It has been suggested that a slightly higher CO_2 concentration could spur plant growth by enhancing the natural process of photosynthesis. However, no one really knows what will happen, and whether the net effect will be good or bad.

Computers have been used in an attempt to figure out exactly what would happen in various parts of the world if the average global temperature were to rise by x°C in y years (x and y being input variables), but this has only increased the confusion, because slight differences in program parameters produce dramatic differences in the predicted effects.

The main problem with global warming may not be the temperature change and its eventual effects, but the possibility that our children and grandchildren might not be able to adapt fast enough without enormous economic and social cost.

There is another possibility that could render all concern about global warming academic: a geological or cosmic catastrophe that would fill the stratosphere with dust for decades, chilling the planet and bringing on an Ice Age. But, of course, it would be silly to do nothing about global warming or other human-caused damage to the environment, based on the hypothesis that an asteroid is eventually going to smash into the earth and put an end to life as we know it, anyway!

PROBLEM 8-5

I've read science fiction books and seen TV shows and movies in which the climate changes catastrophically. For example, the earth's atmosphere goes into "runaway greenhouse mode" and the whole planet ends up like Venus, roasting in oven-like heat, boiling away the oceans and annihilating all life forms. Could this actually happen as a result of human-induced global warming?

SOLUTION 8-5

This is a far-fetched scenario, but it would be arrogant of us to assert that it is absolutely impossible. The idea makes for good stories and movies, though!

Quiz

This is an "open book" quiz. You may refer to the text in this chapter. A good score is 8 correct. Answers are in the back of the book.

1. A warm, arid climate would most likely be found over a land mass at latitude
   (a) 8°N.
   (b) 32°N.
   (c) 56°N.
   (d) 67°N.

2. In the downtown areas of large cities on sunny days, the concrete, asphalt, and other human-made materials commonly cause or contribute to
   (a) intensification of hurricanes.
   (b) frequent lightning.
   (c) heat islands.
   (d) localized droughts.

3. Refer to Fig. 8-9. Suppose conditions are as shown in the map, and it is February. A cold wave in the United States would most likely be taking place
   (a) in the Northern Plains.
   (b) in the Southwest, especially California.
   (c) in the Southeast, especially Florida.
   (d) nowhere.

4. Refer to Fig. 8-9. Suppose conditions are as shown in the map, and it is October. Warm, dry weather in the United States would most likely be taking place
   (a) in the Northern Plains.
   (b) in the Southwest, especially California.
   (c) in the far Northeast.
   (d) nowhere.

5. Refer to Fig. 8-9. Suppose conditions are as shown in the map, and it is December. At this time, we can reasonably expect to observe a wind shift to the northwest along with falling temperatures
   (a) in the South, for example in the Mississippi Delta area.
   (b) in the West, especially in California.
   (c) in the far Northeast.
   (d) nowhere.

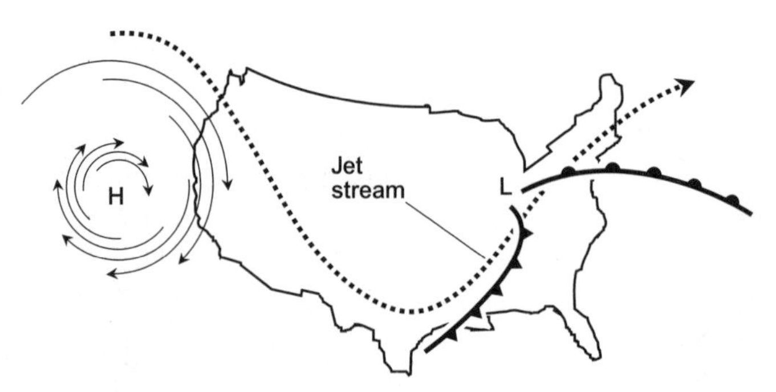

Fig. 8-9. Illustration for Quiz Questions 3 through 5.

6. In the summer, a persistent ridge in the jet stream over the central United States would likely be associated with
   (a) cool weather in the West and a heat wave in the East.
   (b) a heat wave in the West and cool weather in the East.
   (c) heavy rains in the Midwest.
   (d) a heat wave in the Midwest.

7. You can get electrocuted if you wade near a downed utility wire after a flood, even if you don't actually come into contact with the wire, because
   (a) the water, especially dirty flood water, conducts electricity.
   (b) downed utility wires always carry extremely high voltages.
   (c) downed utility wires are almost always "shorted out."
   (d) the power demand is above normal when wires are down.

8. Suppose it is a hot day in the deserts of Nevada. The National Weather Service forecast contains a warning to be careful about overexertion. It does not feel hot to you, but only very warm. At 6:00 P.M., you decide to go out for your 10-km (6-mi) run as usual, even though it is well over body temperature in the shade. This is foolish because
   (a) you could become overheated before you knew it, and collapse with heatstroke.
   (b) the humidity is high, even in the desert, and especially after 4:00 P.M.
   (c) the sun is high in the sky at this hour, and you could get sunburned.
   (d) a sudden rain storm might come up, and you could catch a cold.

9. In January, a deep trough in the jet stream over the Atlantic coast of the United States might result in
   (a) blizzards in California.
   (b) an El Niño event.
   (c) freezing weather in Florida.
   (d) All of the above

10. When the humidity is high, the apparent temperature is lower than the actual temperature on a warm day because
    (a) you sweat more than you would if the humidity were low, and the evaporation helps to cool you off.
    (b) your body temperature cannot rise above normal because the water in the air prevents overheating.
    (c) the water in the air helps to conduct heat away from your body.
    (d) Forget it! The premise is wrong. When the humidity is high, the apparent temperature is higher than the actual temperature on a warm day.

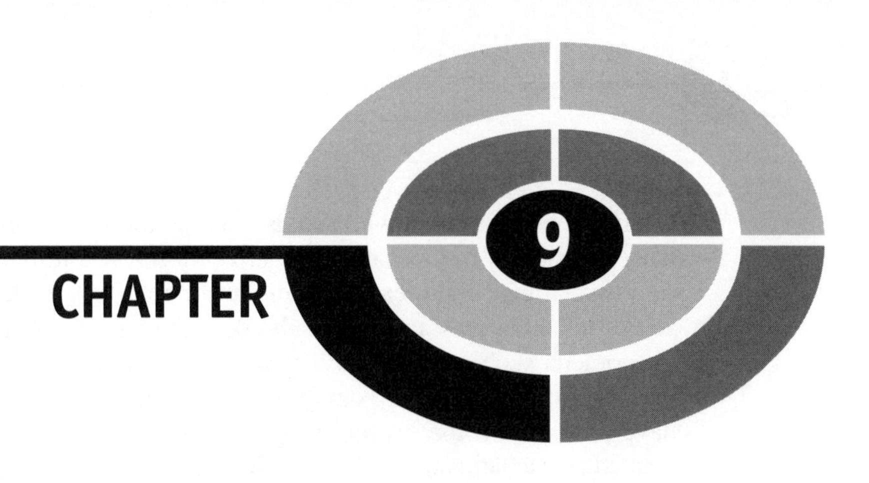

The Past and Future Climate

There was a time when our planet could not support life. There will come a time when the earth will again be uninhabitable. These are phases in the process of planetary birth, life, and death. But there is a variable in the earth's life equation that no other planet, as far as we know, has: humanity!

Early Weather

Some of the earth's original atmosphere was captured from space by gravitation. Some came from the interior of the planet, and was held near the surface by gravitation. The earth's primordial atmosphere contained hydrogen and methane, as well as nitrogen, oxygen, and carbon dioxide in large quantities.

WEATHER IS INEVITABLE

There has been weather on the earth for as long as the planet has had an atmosphere. Any planet with an atmosphere has weather events. Jupiter, for example, has huge, fierce storms with winds comparable to F5 tornadoes on earth. Mars has sand storms that occasionally envelop the whole planet. Astronomers suspect that strange chemical rains might fall on Titan, the largest moon of Saturn.

Before there was any life in the oceans or on the land, there were weather phenomena such as rain, snow, wind, and lightning. Some scientists have suggested that lightning was a catalyst in the formation of *amino acids,* which could have been key in the evolution of the first life forms on earth. Scientists have created amino acids in laboratories by simulating the conditions thought to exist in the atmosphere of our planet long ago.

WEATHER AND LIFE

We all know how rain, sunlight, and heat, or lack of them, affect plant life, but the interaction between plants and weather is more complex. Here are four examples:

- A hurricane blows coconuts off of palm trees as it passes over a small island in the Caribbean. Some of the coconuts land in the ocean and are blown along by the storm until they wash up on a distant shore. The trees, not formerly native to the region, grow and proliferate.
- Birds from the tropics, caught in the eye of a hurricane, are transported thousands of kilometers. This gives rise to new bird species at northerly latitudes.
- A stroke of lightning sets a tree on fire. Other trees nearby are ignited, and a forest fire ensues. Many square kilometers of woodland are scorched. Young trees soon sprout from the blackened land. There is plenty of sun, and the soil is rich in carbon. The trees mature; another fire occurs, and the cycle repeats.
- Wind makes waves on the ocean surface. Some of these have whitecaps caused by tiny air bubbles in the water, generated by the agitation. When raindrops fall on the water surface, the disturbance adds air to the water. Such *aeration* of the water is necessary if fish are to survive, because fish need oxygen.

Glacial and Interglacial Periods

According to recent estimates, the earth is about 4,500,000,000 (4.5×10^9) years old. For millions of years after the first appearance of life on the earth, the climate was moderate over most of the surface. Then the climate became more variable.

This radical change took place recently on a geologic time scale: The severe ice-age climate had become fully established about 1,000,000 (10^6) years ago. There have been numerous glacial advances and retreats since then. The glacial and interglacial periods have been superimposed on a much longer variation in the earth's temperature. Numerous hypotheses have been proposed in an attempt to explain the cause of glacial and interglacial periods. Let's examine a few of them.

EARTH'S ORBIT

The earth orbits the sun at an average distance of about 150,000,000 km (93,000,000 mi), but the actual distance changes during the course of the year. Our planet does not go around the sun in a perfect circle. Instead, the orbit is an *ellipse,* with the sun at one focus (Fig. 9-1).

In July, the separation between the earth and the sun is maximum, approximately 1% greater than the average distance; this is known as *aphelion.* In January, the separation between the earth and the sun is minimum, about 1% less than the average distance; this is called *perihelion.* The sun's brilliance, as "seen" by the earth, is thus about 6% more intense in January than in July! (This fact surprises a lot of people, who think the sun is closest to the earth during summer in the northern hemisphere.)

The earth moves more slowly when it is farther away from the sun, and more rapidly when the distance is less. Thus, we get a few extra days in the northern hemisphere warm season. There are 186 days between (but not including) March 20, which is most frequently the first day of spring, and September 23,

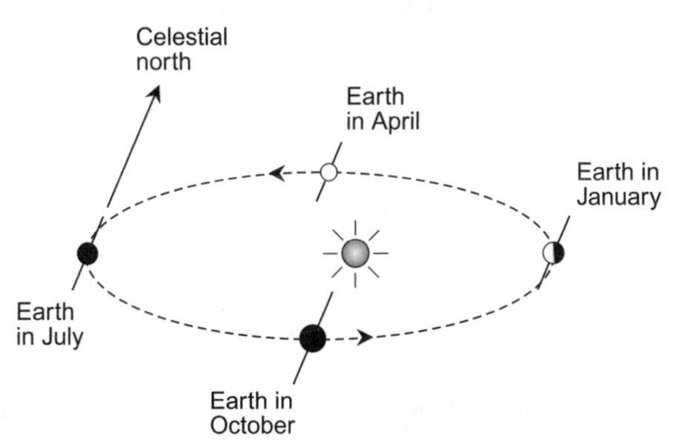

Fig. 9-1. The earth's orbit around the sun is elliptical, with the sun at one focus. (In this drawing the eccentricity of the ellipse is exaggerated.) The earth's rotational axis is tilted with respect to the plane in which the earth revolves around the sun.

which is usually the first day of autumn in a nonleap year. But there are only 177 days between (but not including) September 23 and March 20.

AXIAL TILT

The seasons are caused mainly by the tilt of the earth's axis. The earth's orbit lies in a single plane, known as the *ecliptic plane.* The planetary axis is not perpendicular to the ecliptic plane. If that were the case, the sun would always be above the horizon for 12 hours each day and below the horizon for 12 hours each night, except at the poles, where the sun would follow the horizon, making a complete circle around the compass every 24 hours. In all the inhabited parts of the world, the sun would always follow the *celestial equator,* rising precisely in the east and setting precisely in the west. There would be no seasonal variations in the climate, except for some minor fluctuations caused by the slightly more intense solar radiation at perihelion as compared with aphelion.

ORBITAL PRECESSION

The angle between the ecliptic plane and the earth's axis is about 66.5°. That means the earth is tilted from the perpendicular by 23.5°. As the earth revolves around the sun, the northern and southern hemispheres alternately receive more sunlight, then less, because of this tilt. All planets are tilted in this way, although some (such as Jupiter) are tilted less than our planet, and others (such as Uranus) are tilted more.

The position of the elliptical orbit, relative to this axial tilt and relative to the distant background of stars, slowly changes with time. In a few thousand years, the earth will be closer to the sun in July than in January (Fig. 9-2). This slow change (not noticeable during one human lifetime) is called *precession.* The orientation of the ellipse "spins around in space."

About 6500 years ago, the northern-hemisphere warm season was shorter than the cold season—just the reverse of the present situation. The same thing will happen again about 6500 years from now. Some scientists think that this difference could have a profound effect on the climate, because the northern hemisphere has much more land surface, in proportion to water surface, than does the southern hemisphere. Land retains heat much less readily than water, but land also heats up faster when exposed to solar radiation. The southern hemisphere therefore heats up and cools off more slowly than the northern hemisphere.

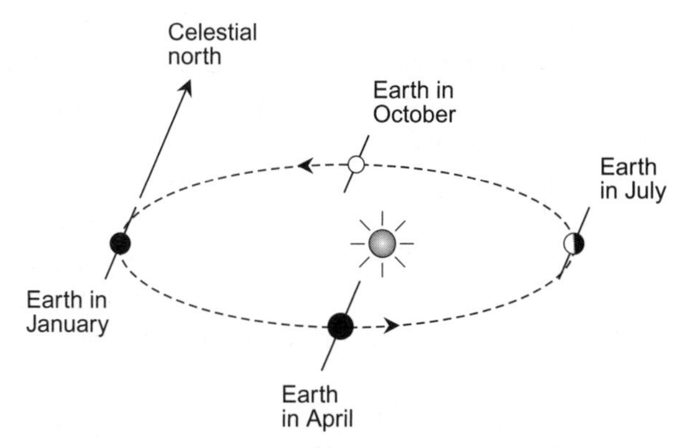

Fig. 9-2. Approximately 6500 years in the future, the orientation of the earth's orbit relative to the stars will be opposite compared with the situation today.

ORBITAL ECCENTRICITY

In addition to the foregoing, the *eccentricity* of the earth's orbit changes with time. Sometimes, as now, the orbit is nearly a perfect circle. But at some times in the past, the earth's orbit has been more elongated, far from being a perfect circle. This will occur again at intervals in the future.

When the earth's orbit is the most elongated, the difference between perihelion and aphelion is such that the amount of solar energy reaching the surface varies by about 25% instead of the current 6%. The *eccentricity cycle* takes about 100,000 years to complete. Astronomers and climatologists alike have suggested that this cycle affects the general climate of the earth.

THE GALACTIC PLANE

Another astronomical phenomenon has been singled out as a possible cause of long-term climatic variability. Our sun orbits around the center of the Milky Way, our galaxy, in much the same way as the earth orbits the sun. The stars in the Milky Way lie mostly, but not entirely, in one plane, and the whole galaxy is shaped something like a discus. Seen from a broadside angle, the galaxy looks like a satellite photograph of a well-developed hurricane (except that the core appears bright rather than dark). Stars are arranged in spiral-shaped arms that bear an uncanny visual similarity to the rain bands in a hurricane. Our solar system, located in one of the spiral arms about 2/3 of the way out from the center to the edge, takes approximately 200,000,000 (2×10^8) years to complete one revolution (Fig. 9-3A). Stars nearer the galactic core revolve around the center of the galaxy faster than our solar system does. Stars farther from the core revolve more slowly than the solar system.

What, you ask, has all this to do with ice ages? As the solar system revolves around the center of the galaxy, it passes in and out of the *galactic plane,* the plane that defines the main disk of the Milky Way. Seen edgewise, our galaxy looks something like the illustration in Fig. 9-3B. Although the stars are concentrated near the galactic plane, their orbits frequently carry them away from this plane. This is true of the sun.

In a spiral galaxy, most of the interstellar gas and dust is concentrated in the galactic plane. Therefore, the sun passes through varying amounts of interstellar matter as it plunges in and out of this plane. Every 100,000,000 (10^8) years or so, the amount of interstellar material around the sun reaches a peak, as the sun passes

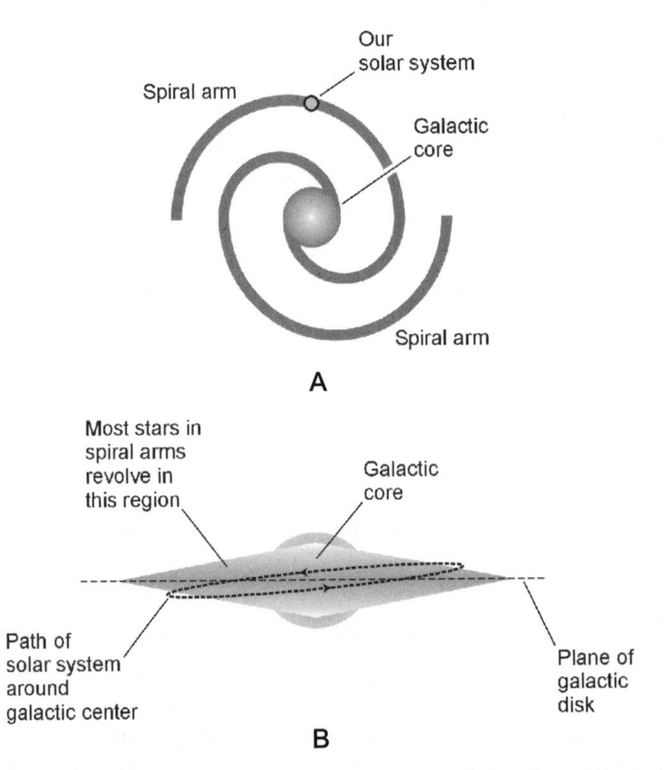

Fig. 9-3. Our solar system is in one of the spiral arms of the galaxy (A). As it orbits the galactic center, the solar system moves in and out of the plane of the galactic disk, which is laced with interstellar gas and dust (B).

through the galactic plane. Midway between these peaks, it reaches a minimum. Some scientists think that the interstellar gas and dust in the galactic plane is sufficient to cause ice ages when the sun is in its midst. The dust, in particular, would dim the sun, in much the same way as dust or smoke in the air can dim the sun.

Perhaps the eccentricity of the earth's orbit, the wobbling of its axis, and the amount of interstellar matter all act on our climate. In that case, there are periods during which all of these factors coincide to favor cooler weather. This, some scientists think, would result in prolonged and extreme ice ages on earth.

SOLAR RADIATION

There is yet another plausible cosmic reason why the earth has had ice ages. Could it be that the sun shines more brightly and generates more heat at some times than at other times? We like to think of our sun as a perfect, stable thing, but we know that it is not. The behavior of the sun is complicated, and although we are learning more about it every day, there is still plenty of mystery. There is evidence that points to a large-term cycle in the intensity of radiation from the sun.

In the center of the sun, where the temperature is millions of degrees Fahrenheit, hydrogen is converted into helium and energy, but it has been suggested that this reaction is not continuous. Suppose it is intermittent, a little like the way a furnace operates as it heats a building? If that is the case, then the nuclear reactions in the sun's core literally cycle on and off. According to this theory, the solar temperature is self-regulating, and the mechanism operates like a thermostat. If the sun starts getting too cool, the furnace (nuclear reaction) is switched on. If things heat up too much, the thermostat stops the reactions. But no thermostat is perfect. If you set the thermostat in your house at a certain level in the winter, the temperature oscillates slightly above and below this value. The highest temperature is reached just as the furnace shuts down, and the coolest conditions occur at the moment the furnace goes on. If the sun has a thermostat-like regulating mechanism, then it should be expected to exhibit a sluggishness of response very much like that in your furnace thermostat. This sluggishness is technically known as *hysteresis,* and is observed in all sorts of devices and systems, both humanmade and natural.

In 1968, experiments were begun in an effort to verify the existence of a strange kind of particle. Theorists produced equations showing that tiny subatomic particles, traveling at the speed of light and capable of penetrating the earth in the way light shines through a pane of glass, are emitted by the sun. These particles, called *neutrinos,* are generated by nuclear fusion reactions. Neutrinos are difficult to detect, because they pass through matter easily. Their penetrating power is so great that most of them pass entirely through the earth unaffected.

Despite their elusive nature, neutrinos were detected using special equipment placed deep underground, so other forms of radiation were blocked by the mass of the earth's crust above. It was assumed that all the detected neutrinos would have to come from extraterrestrial sources, mainly the sun. Far fewer neutrinos were observed than theory predicted. The sun was supposed to be emitting a lot of neutrinos, but it wasn't! Measurement errors were dismissed as a possible cause of these unexpected results; the equipment was tested and found to be working properly. The only plausible solution to this puzzle was that the center of the sun, from which a furious barrage of neutrinos was supposed to be emanating, is far less hot than had been previously thought, and that nuclear fusion—the only possible source of solar energy—is not presently taking place there!

One way out of this conflict between theory and experiment can be called the *solar thermostat theory.* According to this theory, the solar furnace is off at the present time. How long has it been off, and when will it start up again? We cannot be sure about that, but if the solar thermostat theory does represent the facts, it can provide an explanation for the occurrence of ice ages. According to the theory, the sun is cooling off right now. This implies that the earth's climate was once warmer than it is today, and at that time the solar furnace was on. The warmest periods would coincide to those times when the solar furnace was just about to shut down. The theory also suggests that, at some periods in the past, the earth was cooler than it is now; these would be the times when the solar furnace was off, but was just about to start up for another cycle.

MAVERICK STARS

The stars in our galaxy do not all orbit in perfect unison around the center. On a "local" scale, stars move in a random fashion with respect to each other. Besides the sun, the nearest star is about 42,000,000,000,000 (4.2×10^{13}) km, or 26,000,000,000,000 (2.6×10^{13}) mi, distant. That is too far for its gravitational field to affect our solar system. But suppose that, long ago, a maverick star came close enough to perturb the orbits of the planets, including the earth?

A tiny change in the earth's orbit might be enough to change the climate considerably. One problem with the maverick star theory, however, is the fact that the orbital change would be permanent, and the ice ages have always been temporary. However, maverick-star events might be more common than we suspect. Some astronomers have suggested that, scattered throughout the galaxy, there are billions of tiny, dim stars that cannot be seen with optical telescopes. These could, if passing near enough to the solar system, have a subtle effect on the orbits of the planets.

METEORITES

The earth and the other known planets are not the only objects that orbit the sun. Millions of chunks of matter, ranging in size from several hundred kilometers to a fraction of a centimeter across, swarm around our parent star in orbits of all shapes. Some of these orbits are such that collisions occasionally occur between one of the rocks, known as *meteors,* and the earth.

Meteors enter the earth's atmosphere every day. At night, observers can sometimes see the trail of a meteor as it burns up because of friction with the atmosphere. Meteors cause ionization of the upper atmosphere for a short time after passing through, and this ionization can be detected by reflecting radio waves from the "trail." Amateur radio operators often observe this phenomenon. Meteors almost always vaporize completely in the atmosphere, but occasionally a large one makes it to the surface. Then it is known a *meteorite.* Meteorites have been found everywhere in the world.

Every few thousand years, a massive meteorite lands somewhere on our planet. A tremendous explosion occurs, comparable to the detonation of a hydrogen bomb. The impact leaves a crater if it occurs on land. In Arizona, the famous *Meteor Crater,* which is approximately 1.5 km (almost 1 mi) in diameter, was formed by a large meteorite impact several thousand years ago. Other craters have been found and attributed to meteorites.

Every few million years, an extremely large meteor, more accurately called a small *asteroid,* strikes our planet. If the meteor comes down on land, vast quantities of dust are ejected from the point of impact and hurled into the upper atmosphere, where it circulates for years or decades, reducing the amount of solar radiation reaching the surface. The resulting cool period could trigger an ice age.

The meteorite theory is popular among people who favor *catastrophism,* a term used for any theory in which evolutionary change occurs in violent, intermittent, and sudden spurts rather than gradually and steadily. Statistically, the probability of a major meteorite impact is low over short or moderate periods of time. It is not likely, for example, that we will witness a great cosmic collision within the next few years or even the next few decades. However, the earth is billions of years old, and there is excellent evidence to show that there have been several collisions in the earth's lifetime that were violent enough to produce dramatic effects on the global environment. The most catastrophic event of all, which is believed by some astronomers to have caused the formation of the moon, practically destroyed the earth, but it took place before any life existed.

Aside from the dust produced by the actual impact event, a large object striking the earth could trigger an ice age by causing geological disturbances. A violent impact could produce an increase in volcanic activity by upsetting the earth's crust. A long-term rise in volcanism would put more dust into the atmosphere, reducing the amount of solar energy reaching the earth.

PLATE TECTONICS

The continents of the earth, as they appear on a globe or world map today, have not always been the same sizes, had the same shapes, or been in the same geographic locations as they are today. Some geologists believe that approximately 300,000,000 (3×10^8) years ago, all the continents were packed together into a single huge land mass. Over time, this large *crustal plate* broke apart and slid over the viscous, slowly flowing *mantle* in complex ways. These fragments continue to move today, although the motion occurs too gradually to be noticed within a single human lifetime.

In the theory of plate tectonics, the original land mass has been given the name *Pangaea.* It was largely within the southern hemisphere. Most of the earth's surface was ocean, as is the case now—but it was a single, gigantic ocean. The currents in this ocean were vastly different than the currents in the oceans as we know them today, and this gave rise to a much different climatic distribution over the surface. Southern Africa, southern South America, India, and Antarctica were near the south pole, and there is good evidence to show that they were glaciated. As time passed, and the circulation within the earth pushed and pulled on the crust, only Antarctica remained essentially fixed. The other continents moved.

As the continents continue to move (this process has been called "continental drift"), new mountain ranges will rise, old mountains will erode, and the oceans will be redefined. It is reasonable to expect that these changes will produce changes in the global climate, although it is difficult to predict their exact nature and extent.

VOLCANISM

Volcanic eruptions have occurred throughout history. Our lives can be affected, even nowadays, by the eruption of a volcano. Mount Saint Helens provided an excellent example of what can happen when a mountain blows up. It could have been much worse. Whole cities can be destroyed by volcanoes. Pompeii, in the ancient Roman empire, was obliterated forever by a single eruption of Mount Vesuvius. The eruption of Krakatoa in the 1800s produced oceanic shock waves that circled the earth.

There is evidence that the general level of volcanism does not remain constant. If this is true, then a long-term change in volcanic activity could precipitate a long-term change in the earth's climate.

Volcanism could increase or decrease for several reasons. A meteorite impact could cause an increase by "shaking up the earth's crust." Cosmic gravitational disturbances might also have an effect. There has been concern among some people that alignments of the planets, where they all line up on the same side of the sun for a period of a few days, could trigger earthquakes and perhaps also volcanic eruptions. (Extremists cried that this cosmic event would cause California to slide into the Pacific Ocean.) The planets do not align often—there are nine of them, all with different orbital periods—and it is possible that ice ages could have been triggered by this kind of astronomical coincidence. A maverick star, passing unusually close to the solar system, might produce tidal forces sufficient to cause an increase in volcanism. Perhaps, too, volcanism on the earth fluctuates over the eons for reasons we do not yet know.

MOUNTAINS

Before and during the age of the dinosaurs, there were not many mountains on the earth. Thus, mountain-produced weather effects were practically nonexistent.

About 100,000,000 (10^8) years ago, the region roughly corresponding to present-day North America was a large land mass that has been given the name *Laurentia.* Another huge continent at that time, spanning most of the South Atlantic and Indian oceans of today, has been given the name *Gondwanaland.* Near present-day Siberia, a small island continent is believed to have existed; geologists call it *Angara.* Europe and Asia were much smaller than they are today. The continents gradually changed shape and position over the ages. The unrest was accompanied by the formation of mountains. In the United States, the Appalachians were formed during a period that was followed by an ice age known as the *Permian.* At that time, the Appalachians were much higher than they are now; they resembled the modern Rockies. The younger Rockies and Himalayas formed later, along with the Andes, the Pyrenees, and other mountain ranges. The *Pleistocene* ice age followed.

According to one theory, the formation of the earth's mountains is linked to the climate changes that produced the ice ages. Mountain ranges alter the flow of the prevailing surface winds, particularly in North America. Mountain ranges also influence the locations of cold and warm regions on the earth. On the summits of high mountains, snow falls all year long. Snow-capped mountain ranges such as the Andes produce glaciers, and the ice ages have been characterized by massive glaciation. The movement of glaciers is responsible for many of the land features we see today in the temperate and polar regions. The myriad lakes of Minnesota and Wisconsin, for example, were left by retreating glaciers from the most recent ice age.

Some scientists have suggested that the ice ages are the result of general cooling in conjunction with an accumulation of ice in the mountains. When the ice becomes thick enough, it spreads into the lowlands, producing huge glaciers that cover sizable portions of continents. According to this theory, a mountain-building period is going on right now, and this might provide one of the ingredients for another ice age in the future.

THE ROSS ICE SHELF

The continent of Antarctica is almost entirely covered by ice. A vast region of the sea adjacent to the Antarctic coastline is overlaid with ice. A great deal of water is locked up in this "oceanic glacier," which is known as the *Ross Ice Shelf.* If this ice shelf were to break free of the continent, float into the warmer waters to the north, and completely melt, the level of the earth's oceans would rise considerably. This could have a profound effect on the weather everywhere on the earth. At the very least, it would have an economic impact, because it would flood many of the world's low-lying cities and agricultural regions.

There is evidence to suggest that the Ross Ice Shelf accumulated over a period of thousands of years. It is possible that it will someday separate from Antarctica and float into the ocean. A new ice shelf might then start building up, replacing the old one, and when it got too big to be supported by the continent, it too would break away. Suppose this is a natural cycle that has been repeated numerous times in the past? What will happen if the Ross Ice Shelf breaks away from Antarctica and causes a rise in the sea level? Could this trigger an ice age? If this natural cycle actually occurs, how much bigger will the ice shelf get before the next breakaway event happens?

No one can answer the foregoing questions exactly, but one thing can be deduced: If the shelf were to lock up significantly more sea water than it now contains, the ocean level would drop. This would affect the climate in the northern hemisphere if it caused the Bering Strait, the narrow gap between Alaska and Siberia, to close. Ocean currents would then be prevented from flowing between the Pacific and Arctic Oceans. This would have a dramatic effect on the temperature of the water in the North Pacific, perhaps causing the semipermanent high and low pressure regions to move. That would, in turn, cause a shift in the behavior of the jet stream over North America. If the jet stream shifted just a few degrees of latitude toward the south, the climate would be much colder than it is today over the United States and Canada. Conversely, if the jet stream were to move northward, the climate would get much warmer in many places. Precipitation patterns would also be altered.

OSCILLATION

The fact that there have been many ice ages, spaced at intervals of several thousand years, suggests another possible reason for their occurrence. Suppose the earth's ocean-atmosphere mechanism naturally oscillates back and forth between two extremes, a manifestation of *resonance* that is simply inherent in the system? Oscillations are common in natural and humanmade systems; examples include the sun's magnetic field, the back-and-forth motion of the tides, the continuous feedback in a radio transmitter, and the swinging of a pendulum once it has been set in motion. Oscillation represents the alternating, and repeating, storage and release of energy.

Imagine a long, flexible spring, hanging from the ceiling, with a heavy weight at the bottom. If the spring is left alone in a state of equilibrium, the weight does not move. If the weight is displaced upward or downward for any reason, however, the spring-weight system will oscillate (Fig. 9-4). Neglecting the effects of friction, the weight continues to move up and down indefinitely. Suppose an analogous sort of oscillation occurs in the earth's ocean-atmosphere heat engine? If that is the case, the next question is: What could cause the initial imbalance, the "force

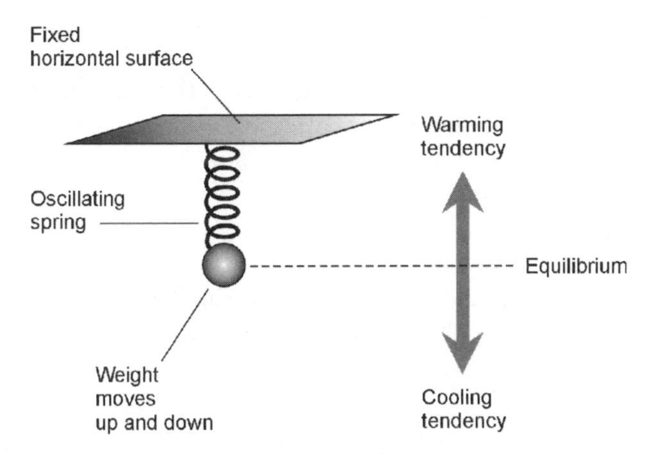

Fig. 9-4. Oscillation occurs in a system once an external force or effect sets it in motion.

that sets the weight in motion"? Possibilities include the precession of the earth's orbit, the wobbling of the planetary axis, a cosmic event such as a meteorite impact, or a sudden change in the amount of volcanic activity in the earth's crust.

PROBLEM 8-1

My friends told me that the sun must be closer to the earth in July than in January, because we all get sunburned much more easily in summer than in winter. But in the above discussion, the claim is made that the sun is closest to the earth in January! I don't believe it. I'm going skiing in the mountains next week, and people have warned me that I had better take sunblock. But I think that is nonsense.

SOLUTION 8-1

The reason people sunburn more easily in summer (all other factors being equal) than in winter is the result of the fact that in the summer, the sun takes a higher course across the sky than it does in the winter. There is less atmosphere between you and the sun in the middle of a July day, as compared with the middle of a January day, because the sun shines down at a steeper angle. The atmosphere acts as an ultraviolet (UV) filter. Thus, the extent of protection it offers depends on the amount of air the sunlight must pass through before it reaches you. In the southern hemisphere, you would sunburn more easily in January than in July. But you had better pay attention to your friends who have told you to use sunblock in the winter! If you go skiing all day and the weather is sunny, even in January, you can get a severe sunburn on your face unless you use UV protection. This is especially true in the mountains, because there is less air between you and the sun when you are at high altitude, as compared with when you are at low altitude. Another factor is the reflection of sunlight from snow. This in effect doubles the effective solar intensity that reaches exposed skin, as compared with the situation when you are on a dark earth or pavement surface.

PROBLEM 9-2

What would the seasons be like if the earth were tilted much more on its axis, say nearly 90°, as is the case with Uranus?

SOLUTION 9-2

The seasons would be far more extreme than they are as we know them today. The length of the day, the angle at which the sun traverses the sky, and the average monthly temperatures would fluctuate to a far greater extent. In fact, over most of the earth's surface, there would be periods of days, weeks, or months (depending on latitude) during which the sun would never rise, and similar periods during which the sun would never set.

The Future: Warming or Cooling?

Some scientists believe that we are in a short-term interglacial period today, and that we are enjoying a mere "blip" in the middle of a long-term ice age. This ice age, according to the theory, has prevailed for millions of years, and can be expected to continue for a long while yet, unless some unforeseen cosmic event or the effects of human activity interfere with the natural cycle.

WHAT IS A GLACIER?

All of the ice ages have been characterized by glaciation of a large part of the northern hemisphere. Even today, glaciers exist in some of the world's mountainous regions, over most of Greenland, and over most of Antarctica.

Glaciers take a long time (in terms of a single human lifespan) to form from snow. For a glacier to develop, the climate must be sufficiently cold so that snow does not melt completely during the summer months. In mountainous areas, this requirement is met primarily because of the altitude, although in Greenland and Antarctica, there is some snow all year round at all altitudes, right down to sea level.

When an ice age commences, snow continues to accumulate over time, more falling than melting each year. This forms a permanent snowpack. The snow is compressed at the bottom of the pack. The individual flakes and crystals lose their identity, and the volume is reduced as the air spaces disappear. Eventually the snow is squeezed so tightly that it becomes ice. This is not like the clear ice that you get in your freezer or on the surface of the pond on your neighbor's farm; instead it is a white, hard mass, like stone. Winds blow snow off the craggy peaks of the mountain and into the passes and crevices, increasing the accumulation there. The ice pack becomes deeper and heavier. Eventually, it becomes a glacier. The glacier proceeds to flow out over the surface, like a thick goo.

We normally think of ice as inflexible and brittle. You cannot, for example, bend an icicle noticeably without breaking it. On a small scale, ice resembles glass, but on a large scale, and over a long period of time, it takes on a much different character. For a moment, ignore the scaling laws of physics, and suppose that you are hundreds of meters tall, and that you perceive time to pass in such a way that a year seems like a minute. With time and space thus altered in scale, glaciers do not appear hard and stiff, but instead they are semiliquid, flowing down mountain slopes and spreading out over huge regions of land. On this scale, ice behaves more like thick molasses than like glass. From this expanded time-space perspective, you would have no trouble envisioning how, given enough snowfall in the polar regions and enough time, glaciers can spread out until they cover much of the land surface of the earth.

If snow accumulates enough to form a glacier and then continues to build up on a mountain as the glacier expands downslope, the land at lower altitudes is gradually covered up by the glacier. Eventually the front of the glacier encounters a climate warm enough to halt its progress, and from that time on, the rate at which the ice melts at the edge of the glacier will exactly balance the forward expansion. If the rate of snowfall increases along with a cooling trend in the climate, the glacier will move farther, and the equilibrium point will move to a lower elevation. If there is still more snow and a still cooler climate, the glacier may move out of the mountain entirely and begin to invade the plains below. This has evidently taken place numerous times in geologic history. Suppose that the most recent glacial invasion of this type was not the last, and that, left untampered with, nature will produce more ice ages in the future?

IT TAKES TIME

Imagine, for a moment, that we are headed for another ice age. Imagine that 20,000 years from now, much of North America, Europe, and northern Asia will once again be covered by ice. What sorts of weather changes will our descendants observe in the coming centuries? The following is a hypothetical description of the climate in North America as an ice age approaches and establishes itself.

There will not be much of a difference in the general weather until about 3000 A.D. Then, early in the 4th Millennium, winters begin to get more severe, especially in the mountains. The last snows of the season in the northern United States occur in late May or early June. The Great Lakes freeze up every year, as they did in the worst winters of the 19th, 20th, and 21st centuries. In the southern United States, snow regularly falls as far south as central Texas and the

Gulf Coast states, with the exception of southern Florida. The regions south of Lake Okeechobee experience regular freezes during the winter, and cold fronts blast down the Florida peninsula from October through June. Similar changes are observed in Europe and Asia. There is little or no change in the equatorial climate, although the monsoons in southern Asia fail more often, producing prolonged and severe droughts.

By 4000 A.D., winters in much of the United States begin in October and last into June, and snow can fall during any month of the year. Frost occurs on at least one day in every month. The general weather in the summertime resembles conditions as they were in 1816, following the eruption of Mount Tambora. A dramatic shift in glacial behavior is noticed; virtually all of the glaciers advance as they gain thickness and mass.

By 5000 A.D., the climate has become so severe that blizzards can occur in the northern United States at all times of the year. Agricultural belts have moved south. The glaciers increase their rate of advance, and northern cities, especially in the continental interior, begin to disappear. In Canada, the glaciers cover some of the high prairies in the west. Palm trees no longer grow naturally anywhere in the continental United States.

The climate continues to grow colder, and the glaciers keep on advancing. By the end of the 10th Millennium (the year 10,000 A.D.), a large part of North America is covered by ice as much as 1.6 km (1 mi) thick, as it was during previous ice ages. Figure 9-5 is a rough map of the United States at this

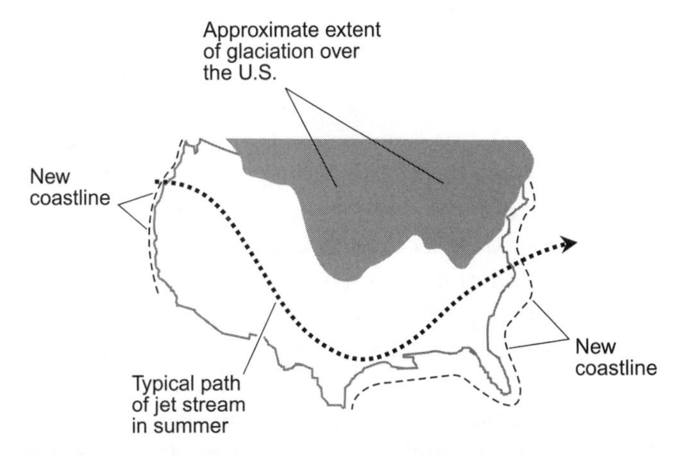

Fig. 9-5. Hypothetical future glaciation of the United States. The regions in gray are under ice. The heavy dashed line shows a typical summertime jet-stream pattern.

time, showing glaciated regions in gray. Temperatures are winterlike all year round in these regions, and cold gales continually blow along the ice fronts.

At last, the equilibrium point is reached, and the rate of melting at the glacial fronts exactly balances the rate of advance. The level of the ocean has fallen, exposing some new land masses. The Bering Strait has been closed by a land bridge. Humans have adapted to the changes, because they have occurred gradually.

CAN WE PREVENT A FUTURE ICE AGE?

Technology has advanced more since the year 1900 than during all of the previous history of humanity. We have learned how to control many aspects of nature, and even, to some extent, the weather. Our control has sometimes been deliberate, and often it has been accidental. We are slowly learning how our activities can affect the climate of our planet. Can we can stop the next ice age from coming, if in fact one is on the way?

Various ways of preventing an ice age have been considered by scientists throughout the world. One notion is that we might attempt to alter the patterns of some of the world's major ocean currents. The Russians once suggested placing a dam across the Bering Strait, preventing the cold Arctic waters from invading the Pacific Ocean. Scientists in the United States and Canada were concerned about the potential climatic effects in North America. Although the climate in North America is variable (as it is everywhere else), conditions are generally good for growing food crops. If the temperature of the water in North Pacific got warmer, the jet stream patterns would change over North America. It is difficult to say what effect this would have on agricultural regions and growing seasons.

Some climatologists think that human activity is already causing effects that might forestall global cooling in the future, and cause global warming instead. This has become a political issue throughout the civilized world, especially in Europe. The burning of fossil fuels produces, among other things, carbon dioxide (CO_2), which helps to trap heat in the atmosphere by means of the greenhouse effect. If another ice age threatens, would we want to deliberately burn great quantities of fossil fuels or other carbon-containing substances all over the world in order to raise the level of CO_2 in the atmosphere?

Another method of warming the earth is to change the *albedo,* or reflectivity, of our planet. Albedo is a term familiar to astronomers. It is expressed as a number between 0 (no reflection) to 1 (total reflection), or a percentage from 0% to

100%. A perfectly black object has an albedo of 0. A perfectly white object has an albedo of 1. Dark objects have low albedo; bright object have high albedo. Clouds reflect much of the energy from the sun, as do snow fields and ice caps. Only a fraction of the sun's light and heat reaches the surface and, of this, most is absorbed. Humans might be able to change either factor to increase the temperature of our planet. The reflectivity of the atmosphere could be changed by reducing the cloud cover; the albedo of the surface could be increased by making it generally darker. (How these things could be done is an open question.) Yet another way to increase the temperature of the earth is by directing more solar energy toward our planet. This could be done by building massive reflectors in space, designed to direct extra sunlight at certain parts of the world.

These days, the mere suggestion of an idea such as "global warming by design" is bound to elicit astonishment. Tampering with the climate in a deliberate way could have long-term effects more devastating for humanity than a naturally occurring ice age. If such an endeavor were successful, the level of the oceans would rise, perhaps enough to inundate major population centers in low-lying coastal areas. Temperature distribution, precipitation patterns, and other factors would be affected in ways that would be very difficult, or impossible, to foretell.

Attempts at deliberate global warming might backfire. Suppose that global warming would not prevent, but in fact contribute to or cause, another ice age? This could take place if precipitation in the polar regions were to dramatically increase, adding to the snowpacks there. Over time, this would produce the very sorts of massive glaciers that have been the signatures of past ice ages. Such phenomena are sometimes manifested in nature as a result of the *pendulum effect.* This effect is based on the fact that the farther a pendulum is pulled offcenter in one direction, the farther it will eventually swing offcenter in the opposite direction. Suppose that the earth has a temperature-regulating mechanism that responds to human global-warming activities by "trying" to cool the planet down?

DEMISE OF THE SUN

There is uncertainty about short-term, medium-term, and long-term climatic trends, but when it comes to the ultimate future, global warming is inevitable. This is guaranteed by solar evolution. Eventually the sun will burn out, but it will not happen for quite awhile. In fact, most scientists believe that the sun will continue to shine for at least 1,000,000,000 (10^9) years into the future, at about the same level of brilliance as it does today. But as the supply of hydrogen runs out, the sun will expand, and its surface will cool off. This has been called the *red-giant* phase, and all stars similar to the sun eventually reach this stage.

The earth will grow warmer, not cooler, as the sun becomes a red giant. This is because the bloated sun will appear much larger in the sky and will send far more energy to the earth than is presently the case. The climate will become intolerably hot; the polar ice caps will melt; wildfires will reduce all plant life to ashes. Sometime during this process, any remaining humans and other mammals will perish. The oceans, lakes, and rivers will boil dry. All living things, even the hardiest bacteria and viruses, will die. The atmosphere will be blown off into space. Some astronomers think that the sun will expand until its radius exceeds the radius of the earth's orbit; the sun will then consume and vaporize the earth.

After the red-giant phase, nuclear reactions within the sun will change in character. Helium atoms will be fused into atoms of carbon, iron, and other elements. The sun will shrink as gravitation once again gains dominance over the internal pressure produced by the energy from the nuclear reactions. Finally, a point in time will be reached at which no further nuclear reactions can take place, and then gravitation will take over. The sun will be crushed into a body of planetary size. Over the ensuing eons, the sun will lose the last of its energy, not unlike a dying ember.

PROBLEM 9-3

I've heard that practically all the earth's glaciers have begun retreating much faster in recent years than they did early in the 20th century. Is this true?

SOLUTION 9-3

Some mountain glaciers have been retreating rapidly and even disappearing, after having remained stable for decades. This suggests that human activities are, in fact, contributing to a general increase in global temperatures.

PROBLEM 9-4

What was the "little ice age"? What caused it?

SOLUTION 9-4

The term "little ice age" refers to a period that lasted from the middle of the 15th century to the middle of the 19th century. During this time, the climate in the northern hemisphere, particularly in Europe, was slightly cooler than it was before about 1450 A.D., and also slightly cooler than it has been since about 1850 A.D. There are several theories that have been put forth to explain this climatic anomaly. One is the idea that the energy output of the sun decreased slightly for about 400 years. Another theory suggests that there was a general increase in volcanic activity around the world, and that the dust from these volcanoes got into the atmosphere and reduced the amount of solar radiation reaching the surface.

Weather and Climate Modification

Weather and climate modification have been attempted, occasionally with success. Sometimes it takes place without direct intent (as with human-induced global warming), and sometimes it is done deliberately. Here are a few examples of deliberate weather modification efforts.

CLOUD SEEDING

Two substances have occasionally been dropped or hurled into clouds in the hope of causing precipitation to fall. Dry ice and silver iodide, in powder form, have been sprayed from airplanes or fired from ground-based cannons. Dry ice chills the air, and silver iodide provides nuclei on which water can condense or freeze. This activity is called *cloud seeding.* It is done in some areas in the hope of increasing rainfall to alleviate drought, and in other places in an attempt to increase snowfall for the benefit of ski resorts.

Cloud seeding has been suggested as a way to reduce the intensity of the winds in the eyewalls of intense hurricanes. During Hurricane Betsy in 1965, seeding was followed by a strange and unexpected course change that brought the storm over the Florida Keys into the Gulf of Mexico when she had been expected to recurve into the open Atlantic. A few people blamed the course change on the seeding, although the scientific evidence points to natural causes. Betsy is not the only example of a hurricane that has suddenly and unexpectedly changed course.

The benefits of cloud seeding must be weighed against the potential for harm. If rain is produced in one place, then another location might be robbed of precipitation. The seeding of hurricanes might have unexpected, potentially harmful consequences, such as increased flooding of low-lying areas.

HAIL CONTROL

Hail is responsible for great damage to crops and personal property each year. As early as about 1900, farmers, in conjunction with scientists, were attempting to control or alter the ways in which thunderstorms produce hail. Because most hail damage is caused by large hail stones, the most common method of hail control has been aimed at getting the hail stones to fall to the surface before they can grow large enough to be destructive.

One early hail control scheme was an attempt to break up the stones, or to get them to fall before they became large. Blanks were fired from cannons to generate shock waves. This method did not prove effective. Later, rockets were shot into storm clouds, with the idea that the disturbance would send the hail earthward while the stones were still small. This did not work either.

More recently, cloud seeding was tried in an effort to interfere with the hail-formation process. Several different countries have used this method, and some believe that it works well enough to be cost effective. The Russians claimed a high rate of hail-control success using artillery to fire shells filled with silver iodide crystals into cumulonimbus clouds.

FOG CONTROL

Clouds form when the air cools to, or below, the dewpoint temperature. Normally this happens at a certain altitude above the surface, but not at the surface itself. Occasionally, however, conditions are such that the base of a cloud is at the surface. This is known as *fog*. It has contributed to many accidents in aviation and highway travel. Fog has plagued mariners ever since the first boat floated on the sea.

Can we get rid of fog, at least on a local scale? Would it be worth the effort and expense? Some airline executives believe that dispersing or burning off of fog would save more money than it would cost by reducing the number of canceled or delayed flights. In harbors, fog dispersal might also prove cost effective. The two primary methods of dealing with fog have been to warm the air above the dewpoint temperature, and to attempt to blow the fog away with aircraft, helicopters, or gigantic fans. Both methods have met with some success. If the fog

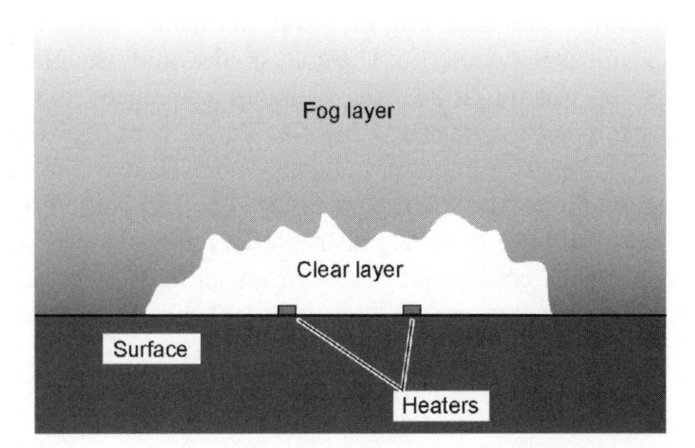

Fig. 9-6. Fog can be cleared near the surface by means of heating units that raise the temperature above the dewpoint.

layer is thin enough (that is, if it does not extend very far above the surface), fans can disperse it by mixing the foggy air with clear air from above. If the fog layer is deep, heaters at ground level can raise the ceiling several meters above the surface (Fig. 9-6).

NUCLEAR WINTER

One of the most frightening climate modification activities, according to scientists, is global nuclear war. In recent years this danger seems to have waned. But a new cold war, perhaps with different geopolitical involvement but equally dangerous potential, could re-evolve at any time.

A large-scale nuclear holocaust would kill millions of people outright, and millions more later as a result of radiation sickness. The explosions would throw large amounts of dust into the air, in much the same way as would a series of huge volcanic eruptions. This material, say the scientists, would cool the whole planet by reducing the amount of solar energy reaching the surface. The effects could last for years or decades. In the extreme, if a "doomsday bomb" is developed, enough fallout could be introduced into the atmosphere to cause dramatic climate change that would persist for centuries or millennia. It would be a climatic jolt unlike anything since the impact of the asteroid that is believed to have been responsible for the extinction of the dinosaurs 65,000,000 (6.5×10^7) years ago.

Following an extreme nuclear war, the survivors would have to endure worldwide freezing as well as radioactivity. The cold spell would take place worldwide. The instigator of the war might bear the worst consequences. Once the cooling began, according to this *nuclear winter hypothesis,* snow packs would melt more slowly than normal. This would increase the albedo of the earth to such an extent that the cooling would persist long after the fallout had settled to the surface. Previously absorbed solar energy would be reflected back into space; cooling would beget cooling. A cool phase, or even an ice age, induced by a global nuclear war might last 100 years, 100 decades, 100 centuries, or longer. No one knows.

PROBLEM 9-5

Suppose a series of hydrogen bombs were exploded in the eye of a severe hurricane while the storm was still out at sea. If the bombs were set off at a high altitude, there would be minimal fallout. Would this work as a method of disrupting a hurricane—literally blowing it apart—before it could strike land and cause a catastrophe?

SOLUTION 9-5

This is an interesting theory, but it is dubious. In less than a minute, a hurricane expends as much energy as the largest known hydrogen bomb. Therefore, it is reasonable to suppose that bombing a hurricane would have little effect. If anything, the heat produced by the bomb might fuel the storm and make it more intense. In addition, some radioactive fallout (however small the amount) would be introduced into the circulation of the storm, and some of this would reach the surface as precipitation nuclei. The result: a radioactive hurricane!

Quiz

This is an "open book" quiz. You may refer to the text in this chapter. A good score is 8 correct. Answers are in the back of the book.

1. Silver iodide is a chemical that is believed to increase the likelihood of
   (a) rain or snow, when it is scattered in clouds.
   (b) global warming, if it accumulates in the atmosphere.
   (c) global cooling, if it accumulated in the oceans.
   (d) the formation of amino acids in bodies of water.

2. Which of the following statements (a), (b), or (c), if any, is false?
   (a) The earth, as a whole, receives slightly more solar energy at perihelion than it does at aphelion.
   (b) The earth, as a whole, receives slightly more solar energy on a day in January than it does on a day in July.
   (c) The earth is closer to the sun in the southern-hemispheric summer than it is in the southern-hemispheric winter.
   (d) All of the above statements (a), (b), and (c) are true.

3. Wave action on the ocean surface
   (a) contributes to expansion of the Ross Ice Shelf, which may cause global cooling and the expansion of the world's forests.
   (b) produces a general drop in sea level, because it causes water to be locked up in glaciers and icebergs.
   (c) helps to aerate the water, and this is important to the survival of marine life.
   (d) is utterly destructive, a fact that has led to a proposal to find ways to reduce it.

4. According to one theory, the earth receives a minimum of energy from the sun at intervals of approximately 100,000,000 (10^8) years because
   (a) the solar system plunges in and out of the galactic plane, which contains interstellar dust and gas that reduces the amount of solar energy that reaches the earth.
   (b) the earth wobbles on its axis, exposing each hemisphere alternately to more solar energy and then less.
   (c) the orbit of the earth precesses around the sun, causing changes in the lengths of the seasons.
   (d) the moon's gravitation disrupts the axial tilt of the earth, causing changes in the intensity of the seasons.

5. Pangaea is the name that has been given to
   (a) a gigantic continent that is believed to have existed on earth eons ago.
   (b) any form of climate change that affects all regions of the world.
   (c) the tendency of earthlike climates to evolve on all planets.
   (d) the effects of weather and climate on the global evolution of life.

6. Neutrinos provide an important clue in solving the puzzle of climatic variability because
   (a) they are associated with the eccentricity of the earth's orbit.
   (b) they come from the galactic plane.
   (c) they can produce greenhouse effect.
   (d) they may help explain how the sun's energy output can vary.

7. Which of the following statements (a), (b), or (c), if any, is false?
   (a) A nuclear winter could increase the albedo of the earth by causing expansion of the snowpack.
   (b) If the Ross Ice Shelf breaks loose from Antarctica, the level of the earth's oceans can be expected to fall.
   (c) If large polar glaciers form, similar to those during the most recent ice age, the level of the earth's oceans can be expected to fall.
   (d) All of the above statements (a), (b), and (c) are true.

8. In the northern hemisphere, summer is a little longer than winter, astronomically speaking, because
   (a) the earth is closer to the sun in the northern-hemisphere summer than in the northern-hemisphere winter.
   (b) the earth is tilted on its axis to a greater extent in the northern-hemisphere summer than in the northern-hemisphere winter.
   (c) the earth travels more slowly in its orbit during the northern-hemisphere summer than in the northern-hemisphere winter.
   (d) All of the above

9. The climate of the earth might be dramatically changed in the long term by
   (a) placing a dam across the Bering Strait.
   (b) doing things that reduce the albedo of the planetary surface.
   (c) increasing the amount of CO_2 in the atmosphere.
   (d) All of the above

10. A theory that describes climate variability in terms of sudden, dramatic events rather than gradual change is known as
    (a) evolutionism.
    (b) reactionism.
    (c) catastrophism.
    (d) the maverick theory.

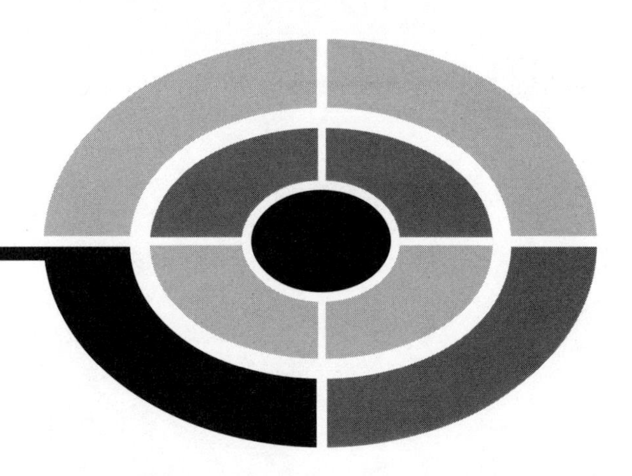

Final Exam

Do not refer to the text when taking this exam. A good score is at least 75 correct. Answers are in the back of the book. It's best to have a friend check your score the first time, so you won't memorize the answers if you want to take the test again.

1. Suppose you want to monitor the wind direction from a convenient location indoors, even when the wind vane itself cannot be seen from that indoor location. What type of device can facilitate this?
   (a) A bi-metal strip
   (b) A selsyn
   (c) A hygrometer
   (d) A compass
   (e) An anemometer

2. Refer to Fig. Exam-1. Suppose you are at Town X. You can expect that the prevailing surface winds at this location come generally from
   (a) the north.
   (b) the south.
   (c) the east.
   (d) the west.
   (e) no particular direction.

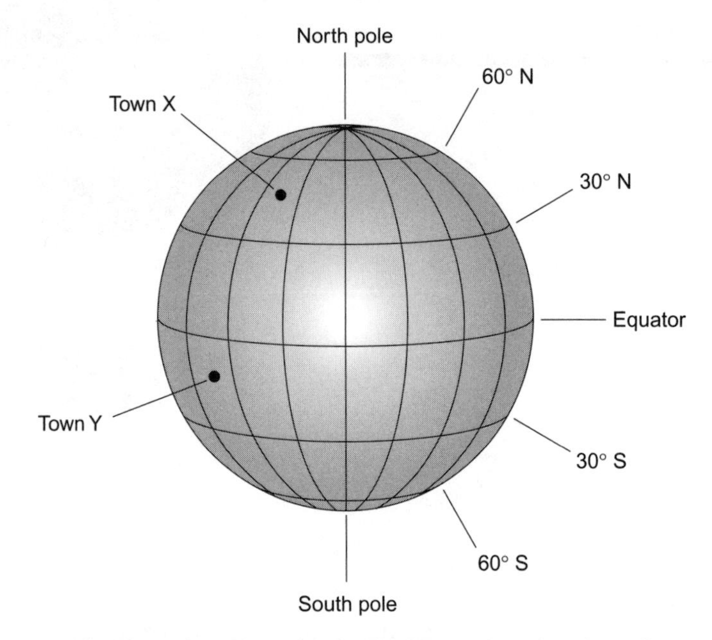

Fig. Exam-1. Illustration for Final Exam Questions 2 and 3.

3. In the situation shown by Fig. Exam-1, suppose you are at Town Y. You can expect that the prevailing surface winds at this location come generally from
   (a) the north.
   (b) the south.
   (c) the east.
   (d) the west.
   (e) no particular direction.

4. Suppose you are reading a paper or scientific article, and you come across a unit symbolized kcal/kg/°C. This is a unit of
   (a) heat energy transfer.
   (b) air pressure.
   (c) thermal density.
   (d) specific heat.
   (e) specific gravity.

5. Of the following cities in the United States, which experiences the most lightning?
   (a) Seattle, Washington
   (b) Portland, Oregon
   (c) Portland, Maine
   (d) Orlando, Florida
   (e) San Diego, California

6. If thunderstorms are near your location, and you hear the civil defense sirens emit a continuous blast for 3 to 5 minutes, it means that
   (a) hail can be expected to fall.
   (b) winds of gale force or stronger are expected.
   (c) a tornado warning has been issued.
   (d) frequent lightning can be expected.
   (e) a severe thunderstorm watch has been issued.

7. Fill in the blank to make the following sentence correct: "Funnel clouds are most likely to develop in the ________ of a tropical cyclone."
   (a) eye
   (b) forward semicircle
   (c) gust front
   (d) ionosphere
   (e) jet stream

8. When air rises because the wind encounters a hill or mountain slope (for example, when the prevailing westerlies blow against the Pacific Coast of the United States), the effect is known as
   (a) orographic lifting.
   (b) gradient lifting.
   (c) frictional lifting.
   (d) cyclonic lifting.
   (e) alpine lifting.

9. The longest winters in the civilized world occur in the interiors of
   (a) Africa and Australia.
   (b) North America, Europe, and Asia.
   (c) South and North America.
   (d) Antarctica and Australia.
   (e) populated deserts.

10. The kilogram per meter cubed is a unit often used to quantify
    (a) the rate of wind divergence.
    (b) the rate of wind convergence.
    (c) the density of a solid.
    (d) the acceleration of gravity.
    (e) the pressure of a gas in a closed container.

11. Fill in the blank to make the following sentence true: “A(n) ________ contains two strips of metal, attached directly and having different coefficients of linear expansion.”
    (a) metal hygrometer
    (b) aneroid barometer
    (c) bi-metal-strip thermometer
    (d) mercury thermometer
    (e) anemometer

12. Imagine a tornado with a wind circulation, and direction of movement, as shown in Fig. Exam-2. The strongest winds in this tornado come from
    (a) the northeast.
    (b) the northwest.
    (c) the southeast.
    (d) the southwest.
    (e) all directions equally.

13. In the tornado diagrammed by Fig. Exam-2, the strongest winds are
    (a) 50 kt.
    (b) 80 kt.
    (c) 110 kt.
    (d) 140 kt.
    (e) 170 kt.

14. The term *mammatocumulus* refers to
    (a) fog over the ocean.
    (b) clouds that seem to hover around mountain peaks.
    (c) clouds within the eyewall of a hurricane.
    (d) the “mother cloud” of a tornado.
    (e) cloud-like formations generated by blowing snow.

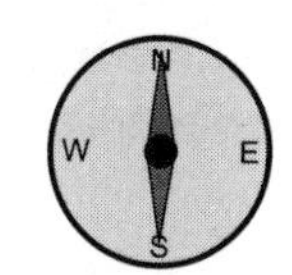

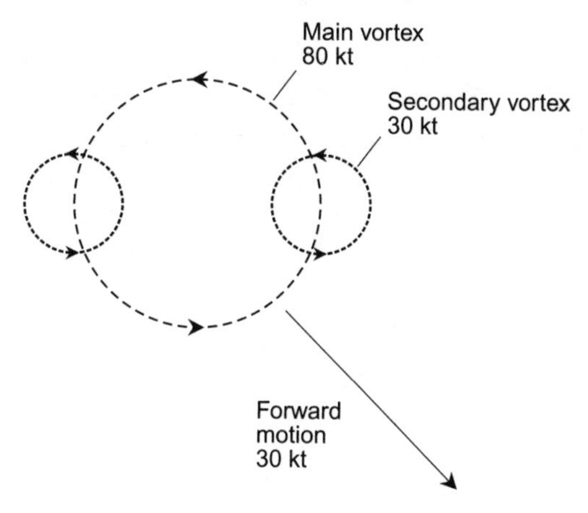

Fig. Exam-2. Illustration for Final Exam Questions 12 and 13.

15. When you attempt to predict the weather for the next 45 minutes by assuming that conditions in general will not change during that time, you employ
    (a) trend forecasting.
    (b) persistence forecasting.
    (c) historical forecasting.
    (d) numerical forecasting.
    (e) synoptic forecasting.

16. When you attempt to predict the weather by extending a defined rate of change into the short-term future, you employ
    (a) trend forecasting.
    (b) persistence forecasting.
    (c) historical forecasting.
    (d) numerical forecasting.
    (e) synoptic forecasting.

17. When you attempt to predict the weather by assembling weather maps from observed and reported data at numerous stations, you employ
    (a) trend forecasting.
    (b) persistence forecasting.
    (c) historical forecasting.
    (d) numerical forecasting.
    (e) synoptic forecasting.

18. When you attempt to predict the weather by observing parcels of air and then using a computer programmed according to laws of physics, to predict what will happen to each parcel at regular intervals into the future, you employ
    (a) trend forecasting.
    (b) persistence forecasting.
    (c) historical forecasting.
    (d) numerical forecasting.
    (e) synoptic forecasting.

19. A squall line is
    (a) a group of large thunderstorms.
    (b) a severe blizzard.
    (c) the eyewall of a hurricane.
    (d) a tornado with straight-line winds.
    (e) a violent dust storm.

20. A typical tornado has winds that
    (a) are somewhat less than gale force.
    (b) can blow in a straight line.
    (c) blow straight down toward the surface.
    (d) blow straight in toward a specific point on the surface.
    (e) None of the above

21. Ultraviolet rays from the sun ionize atoms in the atmosphere. The ionized regions are called the D, E, F1, and F2 layers. The D layer, which is the lowest of the ionized layers, exists in
    (a) the mesosphere.
    (b) the troposphere.
    (c) the jet streams.
    (d) the polar fronts.
    (e) anticyclonic weather systems.

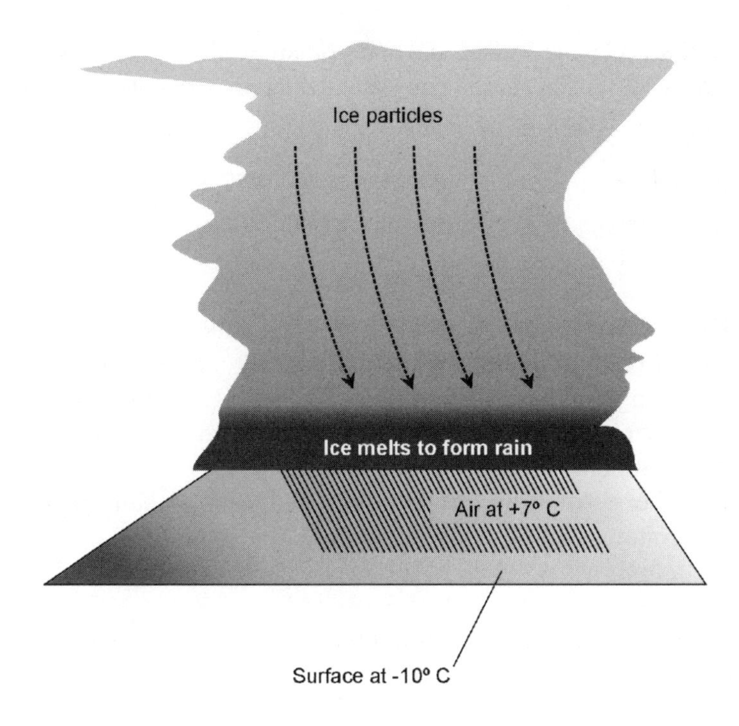

Fig. Exam-3. Illustration for Final Exam Question 22.

22. The scenario illustrated in Fig. Exam-3 is likely to produce
    (a) hail.
    (b) light snow.
    (c) heavy snow.
    (d) freezing rain.
    (e) sleet.

23. Lake-effect snow can be expected on the southern shore of Lake Ontario when the temperature is below freezing, a low-pressure system is passing, and a strong wind is blowing from
    (a) south to north.
    (b) southwest to northeast.
    (c) west to east.
    (d) northwest to southeast.
    (e) southeast to northwest.

24. Suppose you are in Europe, and you are told that the outside temperature is 12°C. This is the equivalent of approximately
    (a) −10°F.
    (b) +54°F.
    (c) +39°F.
    (d) −25°F.
    (e) +72°F.

25. The usual cause of a heat wave in North America is
    (a) a trough in the jet stream that lasts for a long time.
    (b) a ridge in the jet stream that lasts for a long time.
    (c) a series of hurricanes that transfers heat from the tropics to the continent.
    (d) a series of warm fronts associated with a chain of low-pressure systems.
    (e) a long period of unusually wet weather.

26. People who attempt to surf in breakers caused by an approaching hurricane can be endangered by
    (a) littoral currents.
    (b) rip currents.
    (c) undertow.
    (d) the sheer size of the waves.
    (e) All of the above

27. When talking about the frostpoint temperature, another term is often used in its place. What is that term?
    (a) Specific heat
    (b) Sublimation temperature
    (c) Heat of condensation
    (d) Dewpoint temperature
    (e) Thermal equilibrium

28. A supercell is characterized by
    (a) a rotating cumulonimbus cloud.
    (b) the absence of lightning.
    (c) the occurrence of lightning, but only at high altitudes.
    (d) the absence of charge in the atmospheric capacitor.
    (e) zero voltage between clouds and the ground.

29. A potential positive side effect of global warming might be
    (a) a rise in the level of the oceans.
    (b) increased rates of photosynthesis in plants.
    (c) melting of the ice in Antarctica and Greenland.
    (d) desertification of the equatorial regions.
    (e) None of the above

30. Sunlight and high temperatures can contribute to ozone pollution, worsening the air quality in large cities, when
    (a) oxygen atoms in the air group into triplets (O_3) rather than the usual pairs (O_2).
    (b) oxygen molecules in the air are split into single atoms (O) rather than existing as pairs of atoms (O_2).
    (c) oxygen in the air combines with carbon from automobile exhaust and nitrogen that occurs naturally, forming the notorious ozone gas.
    (d) oxygen and nitrogen in the air combine to form nitrates and nitrites, well-known carcinogens that are also known as ozone in gaseous form.
    (e) Forget it! Sunlight and high temperatures tend to reduce air pollution and improve air quality in large cities.

31. Passive solar heating systems are practical in regions where the climate is
    (a) warm most of the time, and sunny most of the time.
    (b) warm most of the time, and sunny some of the time.
    (c) cool most of the time, and sunny most of the time.
    (d) frigid some of the time, and sunny most of the time.
    (e) All of the above

32. The earth revolves most slowly around the sun during the month of
    (a) July.
    (b) September.
    (c) December.
    (d) March.
    (e) Forget it! The earth always revolves around the sun at exactly the same speed.

33. The center of a typical hurricane moves along the storm track at approximately
    (a) 1 to 5 kt.
    (b) 10 to 15 kt.
    (c) 64 kt or more.
    (d) 100 kt or more.
    (e) 150 kt or more.

34. Which, if any, of the following (a), (b), (c), or (d) is a bad idea during an electrical storm?
    (a) If you are caught oudoors, move to a low place away from all structures.
    (b) Try to get into a car or truck if a building is not available for shelter.
    (c) Stand next to a tall tree, so lightning will go through it instead of through you.
    (d) If you are caught in the open, crouch down with your feet close together and your face toward the ground.
    (e) All of the above (a), (b), (c), and (d) are good ideas during an electrical storm.

35. The smoke from a forest fire
    (a) can sometimes be seen on weather radar displays.
    (b) is always associated with cumulonimbus clouds.
    (c) never rises more than 1000 m above the surface.
    (d) is associated with tornadoes.
    (e) produces rain that eventually puts out the fire.

36. The Fujita scale is associated with
    (a) hail storms.
    (b) hurricanes.
    (c) blizzards.
    (d) lightning.
    (e) tornadoes.

37. What is the deadliest phenomenon that occurs in association with an intense hurricane as it makes landfall?
    (a) The tornadoes
    (b) The high winds
    (c) The storm surge
    (d) The heavy rain
    (e) The drop in pressure

38. A station model indicates
    (a) the types of instruments in use at a particular weather station, along with their relative accuracy.
    (b) the theoretical ideal configuration for equipment and processes at a weather station for a given period of time.
    (c) the temperature, extent of cloud cover, wind direction, wind speed, and other information on a weather map for a particular location.
    (d) the barometric pressure at a given weather station, plotted as a function of time as a storm system passes.
    (e) the long-range weather forecast for a particular location, including temperature and precipitation data.

39. Suppose we are told by a reliable source that a ridge in the jet stream will develop, and will be maintained, over the north central United States for the upcoming winter season. If we live in the center of the country (Missouri, for example), what sort of winter can we expect?
    (a) Warmer and drier than normal
    (b) Warmer and wetter than normal
    (c) Colder and drier than normal
    (d) Colder and wetter than normal
    (e) Essentially normal

40. Suppose a heavy, isolated thunderstorm develops over Kansas. A large portion of the storm begins rotating counterclockwise (as viewed from above). This storm has become
    (a) a hurricane.
    (b) an anticyclone.
    (c) an occluded thunderstorm.
    (d) a supercell.
    (e) a squall line.

41. Winds in a typical dust devil
    (a) are rarely stronger than gale force.
    (b) blow at 100 to 150 kt.
    (c) blow at 150 to 200 kt.
    (d) are comparable to those in an F4 tornado.
    (e) are comparable to those in a category 5 hurricane.

42. What is the difference between mass density and weight density?
    (a) Mass density takes the acceleration of gravity into account, but weight density is expressed independently of the acceleration of gravity.
    (b) Weight density takes the acceleration of gravity into account, but mass density is expressed independently of the acceleration of gravity.
    (c) Mass density takes the number of moles of substance into account, but weight density is independent of the number of moles of substance.
    (d) Weight density takes the number of moles of substance into account, but mass density is independent of the number of moles of substance.
    (e) There is no difference.

43. If you are stranded in your car in a blizzard, which of the following is a bad idea?
    (a) Stay with the vehicle as much as possible.
    (b) Run the engine for about 10 minutes per hour.
    (c) Keep the exhaust pipe clear.
    (d) Eat snow as a source of water.
    (e) Turn on the emergency flashers.

44. Fill in the blank to make the following sentence correct: "At a given actual temperature, the windchill factor is the air temperature that would cause the same amount of cooling to exposed flesh as would occur if the same person were walking at a normal pace ________."
    (a) in calm air.
    (b) against the wind.
    (c) with the wind.
    (d) sideways to the wind.
    (e) in direct sunlight.

45. If a small asteroid struck a land mass on the earth's surface, the global climate might become cooler as a result of
    (a) a decrease in the earth's atmospheric pressure.
    (b) an increase in the amount of plant photosynthesis.
    (c) dust thrown into the atmosphere by the impact.
    (d) a decrease in volcanic activity.
    (e) a reduction in the temperature of the earth's core.

46. The severity of a heat wave over a period of time can be quantified in terms of
    (a) kilocalorie-hours.
    (b) energy expenditure.
    (c) cooling degree days.
    (d) heating degree days.
    (e) radiation coefficient.

47. Fill in the blank in the following sentence to make it true: "There is a correlation between ________ and reduced hurricane activity in the Atlantic and Caribbean."
    (a) sunspot activity
    (b) volcanic activity
    (c) the El Niño phenomenon
    (d) the Arctic snowpack
    (e) severe winters in Canada

48. Suppose you live on the oceanfront in the southeastern United States. Which of the following sets of observed phenomena suggests that a hurricane is approaching directly toward you?
    (a) Breakers that increase in size, decrease in frequency, and come from a direction that backs from right to left, along with strong winds that always blow in the direction from which the swells come.
    (b) Breakers that increase in size, decrease in frequency, and come from a direction that veers from left to right, along with strong winds that always blow in the direction from which the swells come.
    (c) Breakers that increase in size, decrease in frequency, and come from a direction that remains constant, along with strong winds that back until they are blowing from your left toward your right as you face in the direction from which the swells come.
    (d) Breakers that increase in size, decrease in frequency, and come from a direction that remains constant, along with strong winds that back until they are blowing directly offshore, regardless of the direction from which the swells come.
    (e) Breakers that increase in size, decrease in frequency, and come from a direction that remains constant, along with strong winds that veer until they are blowing from your right toward your left as you face in the direction from which the swells come.

49. Suppose you live on the oceanfront on the east coast of Australia. Which of the following sets of observed phenomena suggests that a hurricane is approaching directly toward you?
    (a) Breakers that increase in size, decrease in frequency, and come from a direction that backs from right to left, along with strong winds that always blow in the direction from which the swells come.
    (b) Breakers that increase in size, decrease in frequency, and come from a direction that veers from left to right, along with strong winds that always blow in the direction from which the swells come.
    (c) Breakers that increase in size, decrease in frequency, and come from a direction that remains constant, along with strong winds that back until they are blowing from your left toward your right as you face in the direction from which the swells come.
    (d) Breakers that increase in size, decrease in frequency, and come from a direction that remains constant, along with strong winds that back until they are blowing directly offshore, regardless of the direction from which the swells come.
    (e) Breakers that increase in size, decrease in frequency, and come from a direction that remains constant, along with strong winds that veer until they are blowing from your right toward your left as you face in the direction from which the swells come.

50. Suppose 10 g of pure liquid water at +100°C evaporates completely and becomes water vapor at +100°C. In this process, it
    (a) gives up 540 cal of energy.
    (b) gives up 1080 cal of energy.
    (c) gives up 2700 cal of energy.
    (d) gives up 5400 cal of energy.
    (e) does not give up energy, but gains energy.

51. Fig. Exam-4 is a vertical-slice diagram of
    (a) a stationary front.
    (b) a warm front.
    (c) a polar front.
    (d) a tropical front.
    (e) None of the above

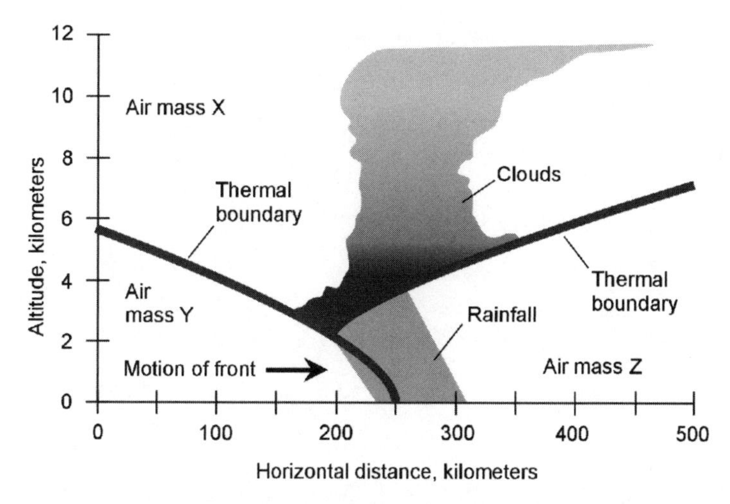

Fig. Exam-4. Illustration for Final Exam Questions 51 and 52.

52. In a situation such as that diagrammed in Fig. Exam-4, the usual case is that
    (a) air mass X is colder than air masses Y or Z.
    (b) air mass Y is colder than air masses X or Z.
    (c) air mass Z is colder than air masses X or Y.
    (d) air masses X, Y, and Z are all the same temperature.
    (e) it is impossible to say anything about the relative temperatures of air masses X, Y, and Z.

53. In an intense tornado, the small-scale wrenching effects that produce spectacular and bizarre damage are caused by
    (a) the mesocyclone.
    (b) the anticyclone.
    (c) hook-shaped echoes.
    (d) suction vortices.
    (e) high barometric pressure.

54. When raindrops fall on the surface of a body of water, the disturbance adds air to the water. This effect is known as
    (a) aeration.
    (b) gasification.
    (c) diffusion.
    (d) ionization.
    (e) irradiation.

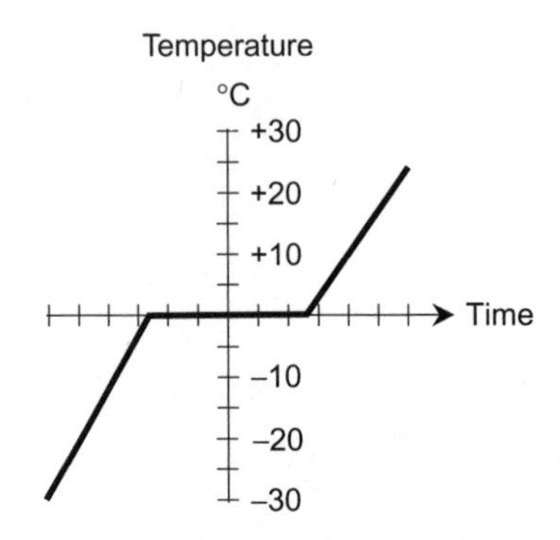

Fig. Exam-5. Illustration for Final Exam Question 55.

55. Fig. Exam-5 is a graph that shows the behavior of
    (a) a sample of air as its pressure increases under the influence of heat energy transfer.
    (b) a sample of liquid water as it evaporates under the influence of heat energy transfer.
    (c) a sample of water ice as it thaws under the influence of heat energy transfer.
    (d) a region of the earth as the sun shines on it and heats it up.
    (e) the heating of a liquid surface under infrared (IR) radiation.

56. The Azores-Bermuda high is a large, persistent system of
    (a) ocean currents in the Pacific.
    (b) storms that track around the Antarctic continent.
    (c) cold fronts that plague the Midwestern United States in winter.
    (d) blizzard-generating storms near the island of Bermuda.
    (e) None of the above

57. Fill in the blank to make the following sentence true: "A cyclonic bend in the jet stream is often accompanied by atmospheric ________, which causes the barometric pressure to drop more than it otherwise would."
    (a) divergence
    (b) regression
    (c) conversion
    (d) compression
    (e) inversion

58. Of the following cloud types, which one most strongly suggests the existence or evolution of a thunderstorm?
    (a) Altostratus
    (b) Cumulostratus
    (c) Cumulus
    (d) Cumulonimbus
    (e) Nimbonimbus

59. Which of the following has been seriously suggested as a possible contributing factor to ice ages on earth?
    (a) The eccentricity of the earth's orbit, and the precession of the orbit with respect to the background of distant stars.
    (b) Periodic changes in the eccentricity of the earth's orbit, so it is sometimes nearly a perfect circle, and sometimes a more elongated ellipse.
    (c) The passage of the solar system in and out of the galactic plane, causing changes in the amount of interstellar dust between the earth and the sun.
    (d) Changes in the actual brightness of the sun, causing more energy to strike the earth during some ages as compared with other ages.
    (e) All of the above

60. Some materials can be considered liquids in the long-term time sense, even though they behave like solids in the short-term time sense, because of their
    (a) density.
    (b) viscosity.
    (c) specific gravity.
    (d) mass.
    (e) pressure.

61. A hurricane watch means that
    (a) hurricane conditions are possible in the watch area within 36 hours.
    (b) hurricane conditions are expected in the watch area within 24 hours.
    (c) hurricanes are likely to form within the next 72 hours.
    (d) conditions are favorable for the formation of hurricanes.
    (e) a hurricane has been sighted.

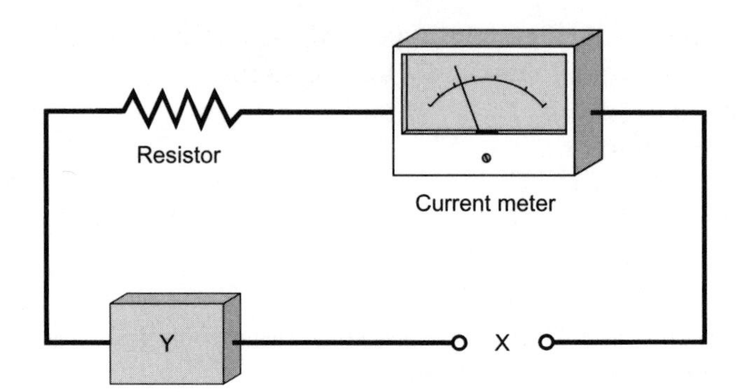

Fig. Exam-6. Illustration for Final Exam Questions 62 and 63.

62. Fig. Exam-6 illustrates an electronic system for measuring relative humidity. What should go in the space marked *X* to make the device work?
    (a) A battery
    (b) A human hair
    (c) A bi-metal strip
    (d) Lithium chloride
    (e) A coil of wire

63. In Fig. Exam-6, what should go in the space marked *Y* to make the device work?
    (a) A battery
    (b) A human hair
    (c) A bi-metal strip
    (d) Lithium chloride
    (e) A coil of wire

64. After the Second World War, American meteorologists started to give names to hurricanes because
    (a) it was made mandatory by an act of Congress after a hurricane wiped out an Air Force base on the island of Hispaniola.
    (b) a hurricane once drove a dog named Daisy so crazy that her master, a meteorologist, decided to name the storm after the dog, and the notion caught on.
    (c) merely numbering them, or giving them other arcane designators, did not make statistical sense because there were too many hurricanes every year.

(d) hurricanes sometimes seem to behave almost as if they are sentient beings, and giving them human names seemed to be appropriate for that reason.
(e) the military, especially the United States Navy, needed a good way to tell them apart and keep track of their progress.

65. The altitude of the bottom of the lowest layer of clouds is called
(a) the cloud ceiling.
(b) the cloud floor.
(c) the cloud level.
(d) the cloud boundary.
(e) the cloud front.

66. Unless you use sunblock, you can get a severe sunburn on your face if you go skiing in the mountains on a sunny winter day because
(a) the earth is closer to the sun in the winter than in the summer, and the result is that sunlight in general is far more intense.
(b) the sun takes a higher course across the sky in the winter than in the summer, so the sun strikes your face at a more direct angle.
(c) at high altitude, there is less air between you and the sun as compared with low altitude.
(d) high winds remove essential oils from your skin, thereby allowing more of the sun's rays to penetrate and do damage.
(e) Forget it! In the winter you don't have to worry about sunburn under any circumstances.

67. Which of the following phenomena is associated with greenhouse effect?
(a) Ozone pollution near the surface reduces the exposure of people in large cities to UV radiation.
(b) Increased amounts of nitrogen in the atmosphere allows more solar radiation to reach the surface.
(c) A decrease in the amount of oxygen in the air slows down the rate of plant photosynthesis.
(d) Longwave IR radiation cannot all penetrate the earth's atmosphere and escape into space, and increased CO_2 exaggerates this effect.
(e) Stronger trade winds produce more severe winters in the polar regions, while at the same time making summers hotter.

68. Particle density can be specified in
    (a) grams per meter cubed.
    (b) grams per centimeter cubed.
    (c) grams per liter.
    (d) kilograms per meter squared.
    (e) moles per meter cubed.

69. A warm westerly or northwesterly wind, blowing down from the mountains, is called
    (a) a gale.
    (b) a mistral.
    (c) an Alberta clipper.
    (d) a Chinook.
    (e) a duster.

70. In numerical forecasting, as the number of iterations of a calculation increases,
    (a) the accuracy of the forecast does not change.
    (b) the accuracy of the forecast increases.
    (c) the error increases.
    (d) the computers become more powerful.
    (e) the parcels of air involved become smaller.

71. Droughts are correlated with
    (a) below-normal temperatures.
    (b) an increase in the frequency of hurricanes.
    (c) above-normal temperatures.
    (d) persistent troughs in the jet stream.
    (e) easterly waves.

72. Which of the following effects (a), (b), (c), or (d), if any, would *definitely not* be observed in conjunction with another ice age?
    (a) Changes in locations of continental coastlines.
    (b) A drop in the general sea level.
    (c) An increase in the size and extent of the polar glaciers.
    (d) A shift in the locations of agricultural regions.
    (e) All of the effects (a), (b), (c), and (d) would likely be observed.

73. A British thermal unit (Btu) is
   (a) the amount of heat energy required to raise the temperature of 1 lb of pure liquid water by 1°F.
   (b) the amount of heat energy required to raise the temperature of 1 g of pure liquid water by 1°F.
   (c) the amount of heat energy required to raise the temperature of 1 oz of pure liquid water by 1°F.
   (d) the amount of heat energy required to raise the temperature of 1 kg of pure liquid water by 1°C.
   (e) the amount of heat energy required to raise the temperature of 1 g of pure liquid water by 1°C.

74. Hypothermia occurs when
   (a) a temperature inversion produces unusually cold air near the surface.
   (b) a body of water freezes from the bottom up, rather than from the top down, as is the normal situation.
   (c) water remains liquid even when the temperature is below freezing.
   (d) a person's body temperature falls significantly below normal.
   (e) the windchill factor drops to less than –40°C (–40°F).

75. As the albedo of an object decreases, that object
   (a) emits more and more UV radiation.
   (b) emits less and less UV radiation.
   (c) reflects more of the light that strikes it.
   (d) reflects less of the light that strikes it.
   (e) None of the above

76. Cloud seeding has been done in an attempt to
   (a) generate hurricanes at sea, where they can help dissipate the excess heat from the tropics without adversely affecting humans who live on land.
   (b) generate tornadoes over unpopulated areas, so they will be less likely to form over populated areas.
   (c) warm up rural areas in the winter by causing cloud cover to increase during the hours of darkness.
   (d) cause existing clouds to produce rain or snow, in an attempt to alleviate drought conditions.
   (e) All of the above

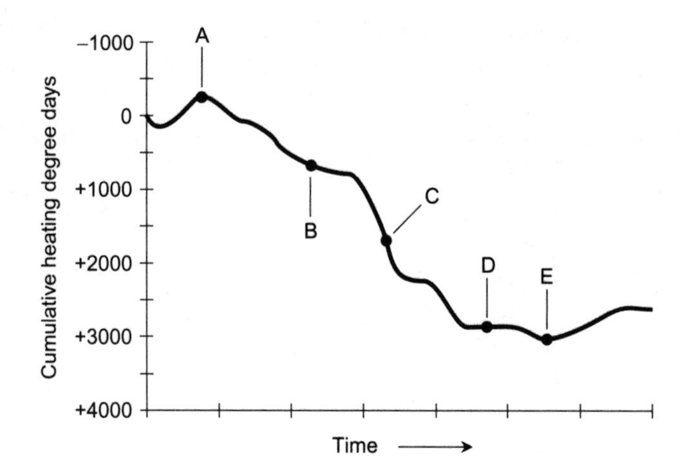

Fig. Exam-7. Illustration for Final Exam Question 77.

77. Fig. Exam-7 illustrates the cumulative heating degree days for a period during the late autumn and winter for a hypothetical location. At which point on the curve would the amount of heating fuel consumed (for example, cubic feet of natural gas per day) be highest?
    (a) Point A
    (b) Point B
    (c) Point C
    (d) Point D
    (e) Point E

78. Which of the following is a bad idea during a flood that threatens to put any part of your house underwater?
    (a) Leave all electrical breakers on to keep power available.
    (b) Do not wade near downed utility wires.
    (c) Try to avoid using a boat if the water is moving.
    (d) Don't drink tap water until you are sure it is safe.
    (e) Move valuables to an upper floor if time allows.

79. The most common method of hail control has been aimed at
    (a) getting the hail stones to fall to the surface before they can grow large enough to be destructive.
    (b) increasing the temperatures at high altitudes, so precipitation cannot freeze to form hail, and will fall as rain instead.

(c) causing massive thunderstorms over unpopulated and nonagricultural areas, in an attempt to release the energy in a storm system before it can produce hail over a populated or agricultural area.
(d) decreasing the temperatures in storm clouds, so hail is more likely to fall as sleet or snow instead.
(e) causing severe thunderstorms to move faster, so they spend less time over any given area, reducing the amount of hail that falls over any particular point on the surface.

80. If a hydrogen bomb were deliberately exploded in the eye of a hurricane, an expected result would be
(a) the movement of the storm away from land.
(b) complete dissipation of the storm within a few minutes to an hour.
(c) a dramatic decrease in the rainfall in the storm.
(d) a dramatic decrease in the wind speed in the storm.
(e) None of the above

81. The extent to which the height, width, or depth of a sample of solid matter changes per degree Celsius is called the
(a) thermal fluctuation ratio.
(b) specific temperature.
(c) specific dimensional thermal variation.
(d) linear differential temperature ratio.
(e) thermal coefficient of linear expansion.

82. In England, cold waves such as those in the north-central United States are unknown, even though England lies further north. This is because
(a) prevailing easterly winds from Europe moderate the English climate.
(b) the large land mass of North America retains more heat.
(c) the warm Pacific current known as La Niña.
(d) the warm Pacific current that creates ridges over the north-central United States.
(e) the warm Atlantic current that originates in the Gulf of Mexico.

83. Neutrinos are generated by nuclear fusion reactions, and have been implicated as a possible factor in the long-term climate of the earth. Which of the following statements about them is true?
(a) They cause radiation sickness to anyone exposed to them.
(b) They cause global warming by heating up the earth's core.
(c) Most of them pass right through the earth.
(d) Few, if any, of them reach the earth's surface.
(e) They cause ionization of the upper atmosphere.

84. Suppose you are looking at data that says it was +12°C at 10:00 a.m. at the weather station in Hoodooburg, +13°C at 11:00 a.m., and +14°C at noon, which is the present moment. From this, you predict that it will be +15°C at 1:00 p.m. at the weather station in Hoodooburg. This is an example of
(a) forecasting by interpolation.
(b) trend forecasting.
(c) synoptic weather prediction.
(d) linear numerical analysis.
(e) geostationary weather prediction.

85. When the high-pressure atmospheric system over the eastern equatorial Pacific strengthens for a period of months or longer, giving rise to strong easterly winds in the tropics over the Pacific, the phenomenon is called
(a) a tropical depression.
(b) La Niña.
(c) a tropical ridge.
(d) El Niño.
(e) an easterly wave.

86. Refer to Fig. Exam-8. This is a diagram of a hypothetical hurricane. Gray curves represent rainbands. The central gray ring represents the eyewall. The central white dot represents the eye. Gray arrows represent surface wind circulation. In which of the following locations might this storm be taking place?
(a) Off the east coast of Australia.
(b) Off the west coast of England.
(c) Off the east coast of the United States.
(d) In the Caribbean.
(e) In the Gulf of Mexico.

87. In the hurricane shown by Fig. Exam-8, at which point, if any (W, X, Y, or Z), is a tornado impossible?
(a) Point W
(b) Point X
(c) Point Y
(d) Point Z
(e) A tornado is possible at any of the four points W, X, Y, or Z.

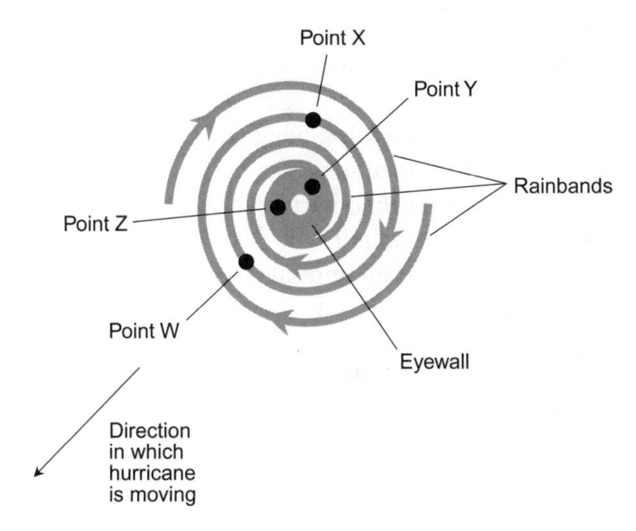

Fig. Exam-8. Illustration for Final Exam Questions 86 and 87.

88. A tropical depression becomes a tropical storm when
    (a) it develops one or more closed isobars.
    (b) its maximum sustained winds reach 34 kt.
    (c) tornadoes begin to form in the periphery.
    (d) it develops a visible eye in satellite images.
    (e) lightning begins to occur.

89. A Doppler radar can reveal rotation within a thunderstorm, a predictor of tornado evolution, by detecting
    (a) changes in the height and density of the clouds.
    (b) changes in the direction from which the echoes are reflected.
    (c) changes in the speed at which the radio waves are returned.
    (d) changes in the frequency of the returned radio waves.
    (e) changes in lightning activity.

90. An Alberta clipper is a
    (a) fast-moving summer thunderstorm that can contain tornadoes.
    (b) high wind from the north that blows over the Rocky Mountains.
    (c) temperature inversion in the Pacific Northwest that produces fog.
    (d) slow-moving, intense, warm high-pressure system over Canada.
    (e) fast-moving winter storm that crosses the northern United States.

91. Suppose we are told by a reliable source that a trough in the jet stream will develop, and will be maintained, over the south central United States for the upcoming winter season. If we live in the center of the country (Illinois, for example), what sort of winter can we expect?
    (a) Warmer and drier than normal.
    (b) Warmer and wetter than normal.
    (c) Colder and drier than normal.
    (d) Colder and wetter than normal.
    (e) Essentially normal.

92. A microburst in a thunderstorm produces
    (a) tornadoes.
    (b) large hail.
    (c) high surface wind gusts.
    (d) localized heavy snow.
    (e) icing of trees and utility wires.

93. As a low-pressure system approaches from the west in the temperate latitudes,
    (a) the barometer reading falls.
    (b) the wind direction may change.
    (c) the wind speed may change.
    (d) the temperature may change.
    (e) All of the above

94. An occlusion occurs when
    (a) a warm front catches up with a cold front.
    (b) a cold front catches up with a warm front.
    (c) a low-pressure system catches up with another low-pressure system.
    (d) a temperate low-pressure system draws a hurricane into itself.
    (e) a high-pressure system overruns a low-pressure system.

95. Fill in the blank to make the following sentence correct: "In a cloud-to-ground lightning stroke, the flashover meets the downward moving ________ a short distance above the ground; a massive electrical discharge follows."
    (a) protons
    (b) air currents
    (c) ice pellets
    (d) ionosphere
    (e) leader

96. Which of the following statements (a), (b), (c), or (d), if any, is false?
    (a) A single tornado may strike a given point on the surface twice.
    (b) Tornadoes occur most frequently in the autumn.
    (c) Tornadoes usually, but not always, form in conjunction with the cyclonic vortex in a large or severe thunderstorm.
    (d) A tornado can occasionally be more than a mile wide at the surface.
    (e) All of the above statements (a), (b), (c), and (d) are true.

97. A weather-data monitoring instrument containing a battery-powered radio transmitter, equipped with a parachute, and released from high-flying aircraft to observe conditions aloft is known as
    (a) a remote monitoring station.
    (b) an anemometer.
    (c) an altimeter.
    (d) a Doppler radiosonde.
    (e) a dropsonde.

98. Hadley cells are
    (a) intense thunderstorms that often contain tornadoes.
    (b) isolated thundershowers within hurricanes.
    (c) regions with positive and negative electric charge that cause lightning.
    (d) low-pressure systems near the poles.
    (e) None of the above

99. Two liquids can gradually mix together in a container at room temperature, even if they don't dissolve in each other or chemically react, because of
    (a) diffusion.
    (b) acceleration.
    (c) their particle density.
    (d) their viscosity.
    (e) their electron shells.

100. In the temperate zone of the northern hemisphere, the prevailing west wind sometimes shifts counterclockwise and blows from the south or southeast. This is called
    (a) a veering wind.
    (b) a steering current.
    (c) a backing wind.
    (d) a cyclonic wind.
    (e) an anticyclonic wind.

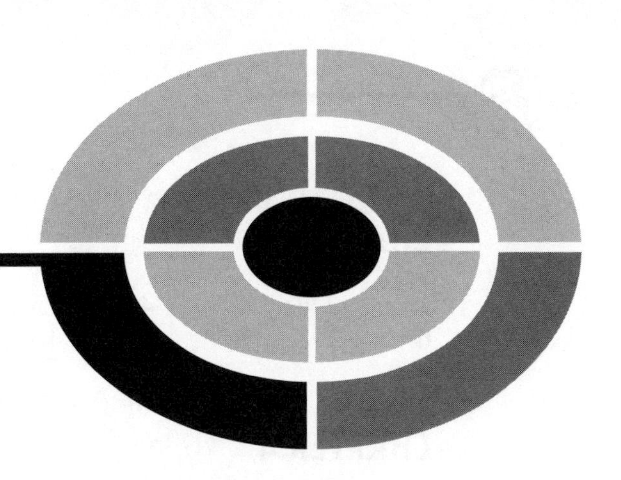

Answers to Quiz and Exam Questions

CHAPTER 1

1. d	2. a	3. b	4. c	5. d
6. b	7. b	8. c	9. a	10. b

CHAPTER 2

1. c	2. c	3. d	4. d	5. a
6. a	7. b	8. c	9. a	10. b

CHAPTER 3

1. a	2. c	3. b	4. a	5. d
6. d	7. c	8. d	9. a	10. c

CHAPTER 4

1. b	2. a	3. b	4. a	5. d
6. c	7. d	8. b	9. d	10. c

CHAPTER 5

1. d	2. c	3. d	4. b	5. a
6. a	7. d	8. d	9. d	10. b

CHAPTER 6

1. c	2. b	3. b	4. c	5. a
6. d	7. b	8. c	9. d	10. d

CHAPTER 7

1. b	2. b	3. c	4. a	5. d
6. c	7. d	8. a	9. b	10. d

CHAPTER 8

1. b	2. c	3. a	4. b	5. a
6. d	7. a	8. a	9. c	10. d

CHAPTER 9

1. a	2. d	3. c	4. a	5. a
6. d	7. b	8. c	9. d	10. c

FINAL EXAM

1. b	2. d	3. c	4. d	5. d
6. c	7. b	8. a	9. b	10. c
11. c	12. b	13. d	14. d	15. b
16. a	17. e	18. d	19. a	20. e
21. a	22. d	23. d	24. b	25. b
26. e	27. d	28. a	29. b	30. a
31. e	32. a	33. b	34. c	35. a
36. e	37. c	38. c	39. a	40. d
41. a	42. b	43. d	44. a	45. c
46. c	47. c	48. c	49. e	50. e
51. e	52. b	53. d	54. a	55. c
56. e	57. a	58. d	59. e	60. b
61. a	62. a	63. d	64. e	65. a
66. c	67. d	68. e	69. d	70. c
71. c	72. e	73. a	74. d	75. d
76. d	77. c	78. a	79. a	80. e
81. e	82. e	83. c	84. b	85. b
86. a	87. e	88. b	89. d	90. e
91. d	92. c	93. e	94. b	95. e
96. b	97. e	98. e	99. a	100. c

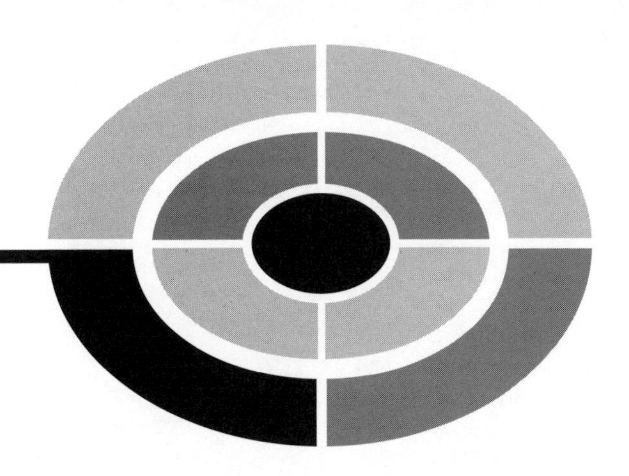

Suggested Additional References

Ackerman, S. A. and J. A. Knox, *Meteorology: Understanding the Atmosphere.* Pacific Grove, CA: Thomson Learning, Inc., 2003.

Cantrell, M. *The Everything Weather Book.* Avon, MA: Adams Media Corp., 2002.

Christopherson, R. W. *Elemental Ecosystems, Fourth edition.* Upper Saddle River, NJ: Pearson Education, Inc., 2004.

Cox, J. D. *Weather for Dummies.* New York, NY: Hungry Minds, Inc., 2000.

Danielson, E. W., J. Levin, and E. Abrams. *Meteorology.* New York, NY: McGraw-Hill, 1998.

Hodgson, M. *Weather Forecasting, Second Edition.* Guilford, CT: The Globe Pequot Press, 1999.

Lutgens, F. K. and E. J. Tarbuck, *The Atmosphere: An Introduction to Meteorology, Ninth Edition.* Upper Saddle River, NJ: Pearson Education, Inc., 2004.

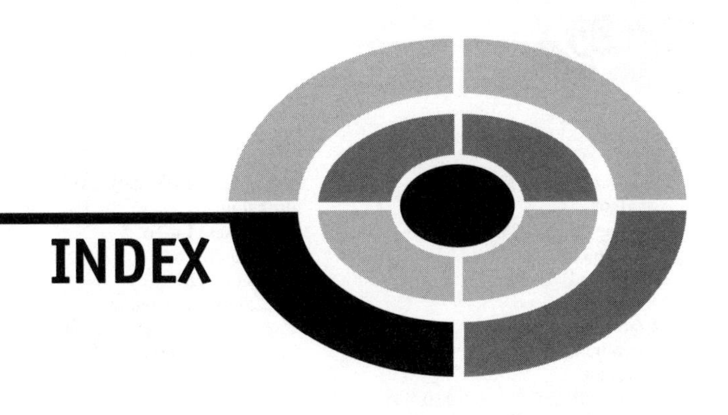

INDEX

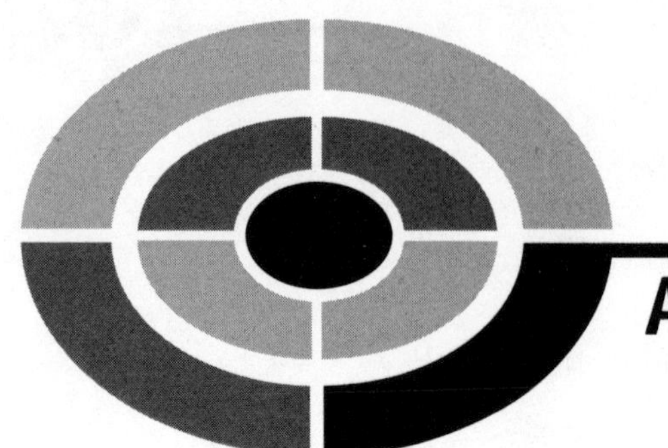

ABOUT THE AUTHOR

Stan Gibilisco is one of McGraw-Hill's most prolific and popular authors. His clear, reader-friendly writing style makes his science books accessible to a wide audience. His background in mathematics, engineering, and research makes him an ideal editor for tutorials and professional handbooks. He is the author of *Teach Yourself Electricity and Electronics, The Illustrated Dictionary of Electronics,* several titles in the McGraw-Hill *Demystified* series, more than 20 other books, and dozens of magazine articles. Booklist named his *McGraw-Hill Encyclopedia of Personal Computing* one of the "Best References of 1996."